WIRELESS SENSOR NETWORKS - INSIGHTS AND INNOVATIONS

Edited by **Philip Sallis**

Wireless Sensor Networks - Insights and Innovations
http://dx.doi.org/10.5772/67388
Edited by Philip Sallis

Contributors

Plácido Rogerio Pinheiro, Alvaro Sobreira Neto, Alexei Aguiar, Pedro Pinheiro, Nuha A. S. Alwan, Zahir Hussain, I.S. Akila, S.V. Manisekaran, R. Venkatesan, Velmani Ramasamy, Prerana Shrivastava, Dr.Sanjay Pokle, Kavi Khedo, Vishwakarma Chikhooreeah, Sabo Miya Hassan, Rosdiazli Ibrahim, Nordin Saad, Vijanth Sagayan Asirvadam, Kishore Bingi, Tran Duc Chung, Noman Shabbir, Syed Rizwan Hassan, Hugo Dinis, Paulo M. Mendes, Akbar Ghobakhlou, Shane Inder

Published by InTech

Janeza Trdine 9, 51000 Rijeka, Croatia

Publishing Process Manager Mirena Calmic
Technical Editor SPi Global
Cover InTech Design team

First published October, 2017
Printed in Croatia
Legal deposit, Croatia: National and University Library in Zagreb

Additional hard copies can be obtained from orders@intechopen.com

Wireless Sensor Networks - Insights and Innovations, Edited by Philip Sallis
p. cm.
Print ISBN 978-953-51-3561-6
Online ISBN 978-953-51-3562-3

For EU product safety concerns:
IN TECH d.o.o., Prolaz Marije Krucifikse Kozulić 3, 51000 Rijeka, Croatia,
info@intechopen.com or visit our website at intechopen.com.

Meet the editor

Dr. Philip Sallis received his PhD degree from the City University (London, England) in 1979. His undergraduate studies in Mathematics and Computer Science were at Victoria University of Wellington, New Zealand. Holding academic positions in London, Sydney and Dunedin, in 1987, Dr. Sallis was appointed the foundation professor of Information Science at the University of Otago. During this time building on the foundation topics of computer science, he built a new curriculum focusing on innovative methods for data analysis and modelling, software engineering, geospatial information processing and knowledge-based systems for automated intelligent decision making.

In 1999, he moved to the then newly formed Auckland University of Technology where until 2009 he was a deputy vice chancellor.

In 2009, Dr. Sallis returned to full-time research, combining his background in Mathematics, Software Engineering and Geocomputation to establish an innovative research centre, which operated with funding domestically in New Zealand and internationally, especially in Chile and Japan for a large collaborative project in the field of wireless sensor networks and their applications for agrometeorology. Now, with his Instrumentation Research Group, Dr. Sallis is collaborating with others in Germany and Sweden to investigate the dynamics of an increasingly popular area of study known as *the Internet of Things* connectivity between devices of all kinds across every sphere of contemporary society.

Dr. Sallis has published extensively over a 40-year scientific career and held senior academic positions in three countries with visiting and adjunct positions in three others. He is a regular conference speaker and is professionally active with groups in business, industry and government agencies. He is a fellow of the Institute of Information Technology Professionals in New Zealand and a member of the Royal Society of New Zealand. He is also a senior member of the IEEE and member of the International Association for Mathematical Geosciences. He regularly reviews journal and conference papers and has served in editorial positions.

Contents

Preface

Wireless sensor networks (WSNs) have emerged as a phenomenon of the twenty-first century with numerous kinds of sensor being developed for specific applications. The origins of WSNs can, however, be traced back to the early days of connectivity between computers and their peripherals. Work with distributed sensor networks is evidenced in the literature during the latter part of the 1970s, continuing in functionality increases in the 1980s and 1990s. As a configuration of independent devices in a data communications network, WSNs are now pre-eminent as working solutions to numerous precision data collection situations where software control of instruments and routing protocols are needed.

A definition for WSNs has been given in a previous InTechOpen publication from 2010 where it was stated 'A WSN consists of spatially distributed sensor nodes. In a WSN, each sensor node is able to independently perform some processing and sensing tasks. ... The usual hardware components of a sensor node include a radio transceiver, an embedded processor, internal and external memories, a power source and one or more sensors'. [Qinghua Wang and Ilangko Balasingham, Wireless Sensor Networks – An Introduction, InTechOpen, DOI: 10.5772/13225]

With the advent of the 'Internet of Things' (IoT) concept, WSN configurations provide the platform for all manner of sensors to potentially communicate with devices/sensors and another in real time, using the Internet for connectivity and 'The Cloud' for data storage and exchange. In this, with all the attendant data and system security issues and contemporary methods for mining, filtering and visualizing information, there is a conceptual description of what and how modern technology is facilitating our world of communication, saturated by collected data that is interpreted and reinterpreted as meaningful information.

In this book, the authors have chosen a selection of specific topics relating to WSNs: their design, development, implementation and function. Some operating topics are addressed such as power management, data interchange protocols, instrument reliability and system security. Other topics are more application oriented, where particular hardware and software configurations are described to deliver system solutions for specific needs. All are clearly written with considerable detail relating to each of the issues addressed by the authors.

Each of the chapters provides a rationale for the topic being covered and some general WSN details where appropriate. The citations used in the chapters are comprehensively referred to, which adds depth to the information being presented.

Philip Sallis
Professor in Computer Science
Telemetric Research Centre
School of Engineering, Computer and Mathematical Sciences
Auckland University of Technology, New Zealand

Section 1

WSN Principles & Practice

Chapter 1

Mobile Wireless Sensor Networks: An Overview

Velmani Ramasamy

Additional information is available at the end of the chapter

http://dx.doi.org/10.5772/intechopen.70592

Abstract

Mobile wireless sensor networks (MWSNs) have emerged and shifted the focus from the typical static wireless sensor networks to networks with mobile sensor nodes that are capable to sense the various types of events. Also, they can change their position frequently in a specific sensing area. The applications of the MWSNs can be widely divided into time-driven, event-driven, on-demand and tracking based applications. Mobile sensor node architecture, residual energy utilization, mobility, topology, scalability, localization, data collection routing, Quality of Service (QoS), etc., are the key factors to design an energy efficient MWSNs for some specific purpose. This chapter deals with an overview of the MWSNs and a few significant phenomena to design an energy efficient MWSNs to the large-scale environment.

Keywords: mobility, mobile sensor node, routing, topology, data collection, MWSNs

1. Introduction

Mobile wireless sensor networks (MWSNs) play a vital role in today's real world applications in which the sensor nodes are mobile. MWSNs are much more versatile than static WSNs as the sensor nodes can be deployed in any scenario and cope with rapid topology changes. Mobile sensor nodes consist of a microcontroller, various sensors (i.e., light, temperature, humidity, pressure, mobility, etc.), a radio transceiver, and that is powered by a battery [1]. The major applications of MWSNs are economics, environmental monitoring, mining, meteorology, seismic monitoring, acoustic detection, health care applications, process monitoring, infrastructure protection, context aware computing, undersea navigation, smart spaces, inventory tracking and tactical military surveillance [2]. There are two sets of challenges to MWSNs; hardware and environment. The main hardware constraints are limited battery power and low-cost requirements. i.e., the mobile sensor nodes should be energy efficient, low complexity algorithms required for microcontrollers and use of only a simplex radio [3].

The mobility models to define the movements towards/away the sensor nodes, and how the mobile sensor nodes location, velocity and acceleration change over time, also predicts the future node positions.

In MWSNs, the major environmental factors are the shared medium and varying topology. The shared medium denotes that channel access must be regulated in some way. Hence, the network topology plays a significant role in routing protocol design and also decides the transmission path of data packets to reach the desired destination [4, 5]. While the sensor nodes on mobility, the performances of the network topologies such as flat/unstructured, chain, tree and cluster topologies are inadequate for large-scale MWSNs. To solve these kinds of issues a hybrid network topology is the best option for large-scale environments. Furthermore, the hybrid topology plays a significant role in data collection as well as the network performance is also good. Also, the routing protocol decides the efficient and reliable data transmission path. Therefore, this chapter deals with the various types of WSN as well as the design challenges, mobile sensor node architecture, mobility entity and mobility models, network topology and several routing protocols for MWSNs [6, 7].

2. Types of WSNs

Usually, the sensor nodes are deployed on land, underground and under water environments and that forms a WSN. Based on the sensor nodes deployment, a sensor network faces different challenges and constraints. Types of the WSNs are terrestrial, multimedia, underground, multi-media and mobile WSNs. In this chapter, we are discussing the overview of the mobile WSNs. According to the resources of the sensor nodes on an MWSN, it can be classified into homogeneous and heterogeneous MWSNs [3]. Homogeneous MWSN consists of identical mobile sensor nodes and they may have unique properties. But, heterogeneous MWSN consists of a number of mobile sensor nodes with different abilities in node property such as battery power, memory size, computing power, sensing range, transmission range, and mobility, etc. Also, the nodes deployment of heterogeneous MWSN is more complex than homogeneous MWSN [8, 9].

2.1. Why are mobile nodes considered in WSNs?

Kay Romer and Friedemann Mattern investigated the design space of the wireless sensor networks and suggested many applications such as bird observation on great duck island, zebranet, cattle herding, bathymetry, glacier monitoring, cold chain management, ocean water monitoring, grape monitoring, power monitoring, rescue of avalanche victims, vital sign monitoring, tracking military vehicles, parts assembly, self-healing mine field, and sniper localization. Among 15 different applications, 10 applications are purely mobile and one of them is partially mobile. Therefore, mobile sensor nodes play an important role in humans real world applications [10, 11].

3. Design challenges of MWSNs

The major design challenges to the MWSNs are hardware cost, system architecture, deployment, memory and battery size, processing speed, dynamic topology, sensor node/sink mobility, coverage, energy consumption, protocol design, scalability, localization, data/node centric, network heterogeneity, node failure, QoS, data fusion/redundancy, self-configuration, cross layer design, balanced traffic, fault tolerance, wireless connectivity, programmability and security [12–14].

4. Mobile sensor node architecture

Usually, the sensor nodes are designed with one or more sensors (i.e., temperature, light, humidity, moisture, pressure, luminosity, proximity, etc.), microcontroller, external memory, radio transceiver, analog to digital converter (ADC), antenna and battery. Again, the nodes are limited on-board storage, battery power, processing and radio capacity due to their small size [15]. However, the mobile sensor node architecture is almost similar to the normal sensor node. But, some additional units are considered on mobile sensor nodes such as localization/position finders, mobilizer, and power generator. The architecture of the mobile sensor node is shown in **Figure 1**. The location or position finder unit is used to identify the position of the sensor node and the mobilizer provides mobility for a sensor node. The power generator unit is responsible to generate a power for fulfilling further energy requirements of the sensor node by applying any specific techniques such as the solar cell.

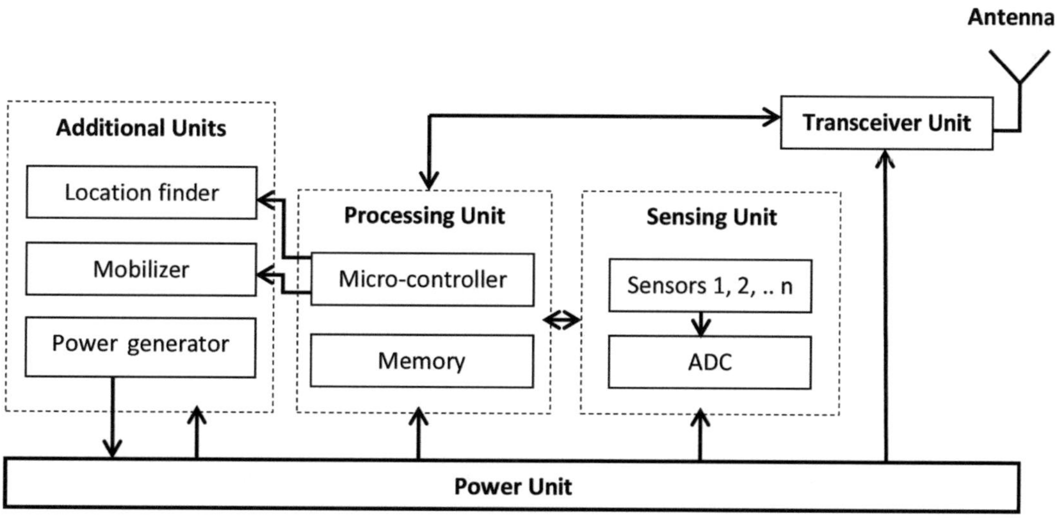

Figure 1. Architecture of the mobile sensor node.

5. Mobility entity

Nowadays, the researchers considering the MWSNs for large scale applications and that consists of a large number of sensor nodes and sink nodes. Here, the mobility can be applied to the sensor nodes or sinks depending on the application requirements.

5.1. Why are mobility models considered in MWSNs?

Usually, the MWSN is a self-configuring and a self-healing network which consists of mobile sensor nodes connected wirelessly to form an arbitrary topology. A good coverage network ensures the reliable communication, higher network connectivity, lower energy consumption and consequently longer lifetime of sensor nodes [16]. Mobility models characterize the mobile sensor nodes movement patterns, i.e., the different behaviors of the nodes. Several mobility models have been considered in MWSNs to set the mobility of mobile nodes. Here, the sensor nodes movements are considered as an independent or dependent of each other respectively.

5.2. Mobility models

The mobility modeled to describe the movement towards/away mobile sensor nodes, and how the mobile sensor nodes location, velocity and acceleration change over time. Mobility models are frequently used for simulation purposes and that is used to investigate the new communication or navigation techniques. Mobility management schemes for mobile wireless sensor networks describe the use of mobility models to predict the future positions of the sensor node [4].

In mobility modeling, the sensor nodes movement can be defined using both analytical model and simulation model. Here, the input of the analytical mobility model simplifies the assumptions of the movement behaviors of the sensor nodes. Also, the analytical model will provide the performance parameters for the simple cases of mathematical calculations. These models can offer the functioning constraints for simple events through scientific calculations. In contrast, simulation models are considered as a well-defined realistic mobility scenario, and that derives the constructive solutions to more complicated cases. Again, the mobility models accurately represent the mobile sensor nodes in the MWSN which is the key to examine the designed protocol is beneficial in a specific type of mobile scenario [17].

The modeling of the mobility patterns can be considered into (a) *trace models*: a deterministic mobility pattern of real-life systems; (b) *syntactic models*: represents the movements towards/away mobile sensor nodes realistically. It can be classified into individual mobile movements and group mobile movements. The mobility models can also be considered by mobility patterns and histories such as directional, random and habitual mobility models. Various mobility models are classified into four major categories which are based on their specific characteristics and that includes random models, models on temporal dependency, models on spatial dependency and models on geographical restrictions. **Figure 2** shows the classification of mobility models based on their areas.

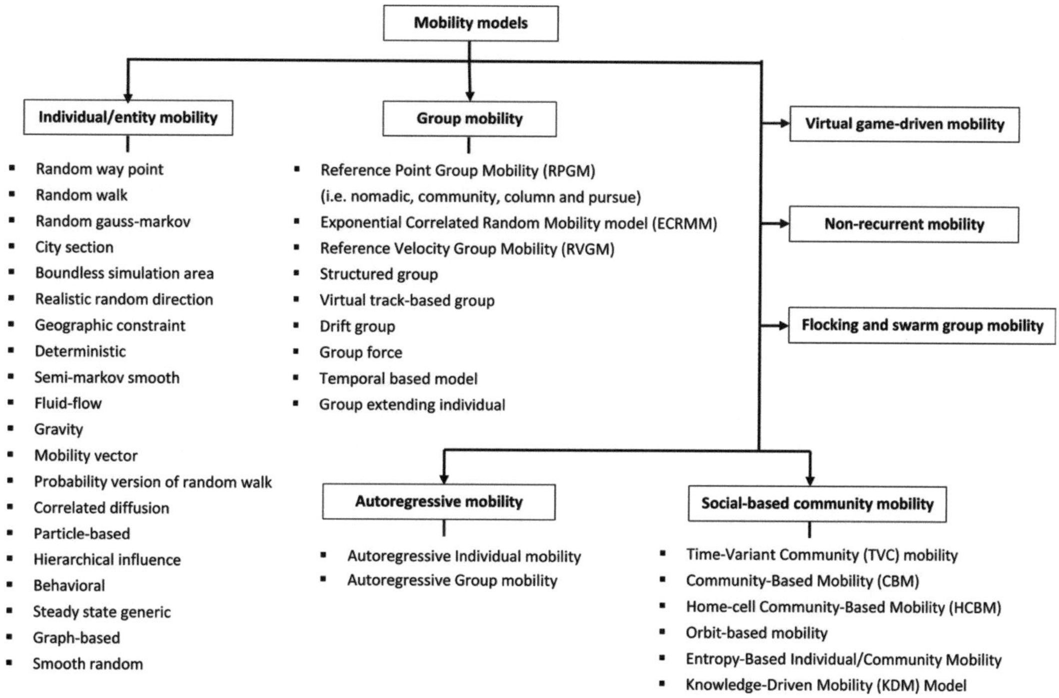

Figure 2. Classification of mobility models based on their areas.

The mobility models can also be classified into the following areas [18]:

- **Individual/entity mobility models:** represents the mobility pattern of the individual mobile node. e.g. random waypoint, random walk, random Gauss-Markov, city section, boundless simulation area, realistic random direction, probability version of the random walk, geographic constraint, deterministic, semi-Markov smooth, fluid-flow, gravity, mobility vector, correlated diffusion, particle-based, hierarchical influence, behavioral, steady state generic, graph-based and smooth random mobility models.
- **Group mobility models:** co-operative groups movement towards/away the mobile nodes acts in synchrony as a group. Here, the movements of the mobile nodes are not independent of each other. Also, the function defines the group behavior or the mobile nodes are somehow connected with a target or a group leader. Many group mobility models are exist such as reference point group mobility (RPGM) (i.e., nomadic community, column and pursue), exponential correlated random mobility model (ECRMM), reference velocity group mobility (RVGM), structured group, virtual track-based group, drift group, group force, temporal based model and group extending individual mobility models.
- **Autoregressive mobility models:** the mobility pattern of individual sensor node/group of sensor nodes correlating the mobility status and that may consist of position, velocity, and acceleration at consecutive time instants. e.g. autoregressive individual mobility model and autoregressive group mobility model.

- **Flocking and swarm group mobility models:** a collective action of a massive number of cooperating mobile agents with a mutual group objective. Examples of the agents are ants, bees, fish, birds, penguins, and crowds. In a flocking mobility model, a coordinated movement task performed by dynamic mobile nodes or mobile agents over self-organized networks of nature. Self-organizing features of the flocks/groups/schools provide a deeper insight into designing MWSNs. Since random walk (RW) and random way point (RWP) models are not suitable for realistic environments, swarm group mobility model is introduced to generate the realistic movements of living organisms or objects led by living organisms by psychological behaviors, physics, and mimicking perceptions is explained.
- **Virtual game-driven mobility models:** based on the user requirements, an individual/group of mobile sensor nodes are characterized from the real time to virtual agents cooperating with each other groups of mobile users. It models the real-world characteristics of the user, group, communication, and the environment. Here, a virtual world is used for the simulation of mobility and that includes all features of mobility models.
- **Non-recurrent mobility models:** nodes mobility on the unknown way of unrepeatable the previous patterns. Let, the mobile nodes can be moving data objects which continuously changing its topology. Here, kinetic data structures (KDS) is considered to capture a continuous moving data objects in an information database when the mobility of the data object can be defined as a polynomial of time to collect the non-recurrent mobility pattern of the object. KDS is fully or partially imaginable. Further, random mobility is captured by soft kinetic data structures (SKDS). KDS and SKDS maintain an approximate geometric structure which is updated by property testing and reformation.
- **Social-based community mobility model:** each mobile sensor node is considered as a member of a cluster of a community whereas different communities may be a part of an overall society. The model must capture the non-homogeneous activities in both space and time normally known in certainty with mathematically tractable. e.g. time-variant community (TVC) mobility, community-based mobility (CBM), home-cell community-based mobility (HCBM), orbit-based mobility, entropy-based individual/community mobility and knowledge-driven mobility (KDM) model.

6. Network topology

The network topology plays an important role in MWSNs to transfer the data onto the mobile sensor nodes to the sink/base station. Then, the sink and the remote user/server are connected by the internet. The effectiveness of large scale mobile wireless sensor networks purely to depend on the data collection or topology management scheme. Therefore, the topology provides a guaranteed reliable network and better QoS in terms of mobility, traffic, end-to-end connection, etc. In addition, topologies in MWSNs define the dimension of the sensor node group, manage the addition of new members of a group and deal with the withdrawal of members that leave the group. With considering such aspects in network topology may provide an efficient data collection of low energy utilization and form superior MWSN. The existing network topologies of WSNs are flat/unstructured, tree, cluster, chain, and hybrid. Various network topologies are followed to achieve the maximum data collection and network performance, which depends upon the nature of the MWSNs. Depend upon the nature of the network, various network topologies are followed to obtain the maximum data collection.

7. Routing protocols for MWSNs

The routing approaches to the MWSNs can be centralized, distributed or hybrid. An efficient and reliable routing protocol design for MWSNs considers the network topology, sensor node mobility, energy consumption, network coverage, data transmission methods, QoS, connectivity, data aggregation, sensor node and communication link heterogeneity, scalability, and security. The existing routing protocols are grouped based on their routing structures such as flat, hierarchical and location based routing protocols. Again, the hierarchical based routing can be broadly categorized into classical and optimized hierarchical based routing. Furthermore, the path establishment discovers the route from the source to the destination which follows the proactive, reactive, and hybrid based routing [5, 19, 20]. **Figure 3** shows the taxonomy of the mobile wireless sensor networks [21–26].

7.1. LEACH variants

LEACH is one of the most popular dynamic clustering algorithms for the hierarchical routing based sensor networks, which is absolutely designed for the distributed environment and there is no global knowledge required about the network. The sensor node uses the received signal strength as well as the threshold values to select the cluster head and that forms a cluster. Here, the round or topology updates interval are considered for the data transmission, which is divided into fixed time intervals with equal length. Each sensor node on a network has an equal probability to act as a cluster head by selecting a random number between 0 and 1, and therefore, the sensor nodes die slowly [27]. The total network operations are considered as a round. Each round consists of a setup phase (i.e., forms a cluster and makes a multi-hop communication between the cluster head and the sink) and a steady state phase (i.e., transfer the data of the cluster members to the sink through the cluster heads). The data transmission phase of the LEACH protocol consists of the intra-cluster and the inter-cluster communications. In an intra-cluster communication, the cluster head collects data of the cluster members and aggregate that data instantly. After the completion of the intra-cluster communication, inter-cluster communication is started to forward the data of the cluster heads to the sink.

Even dynamic clustering improves the lifetime of the network; the static LEACH is not suitable for the large scale mobile wireless sensor networks; therefore, LEACH with mobility must be considered for the dynamic networks. The LEACH variants such as T-LEACH, mobile LEACH, and LEACH-mobile-enhanced are considered for the mobile sensor networks [29–32].

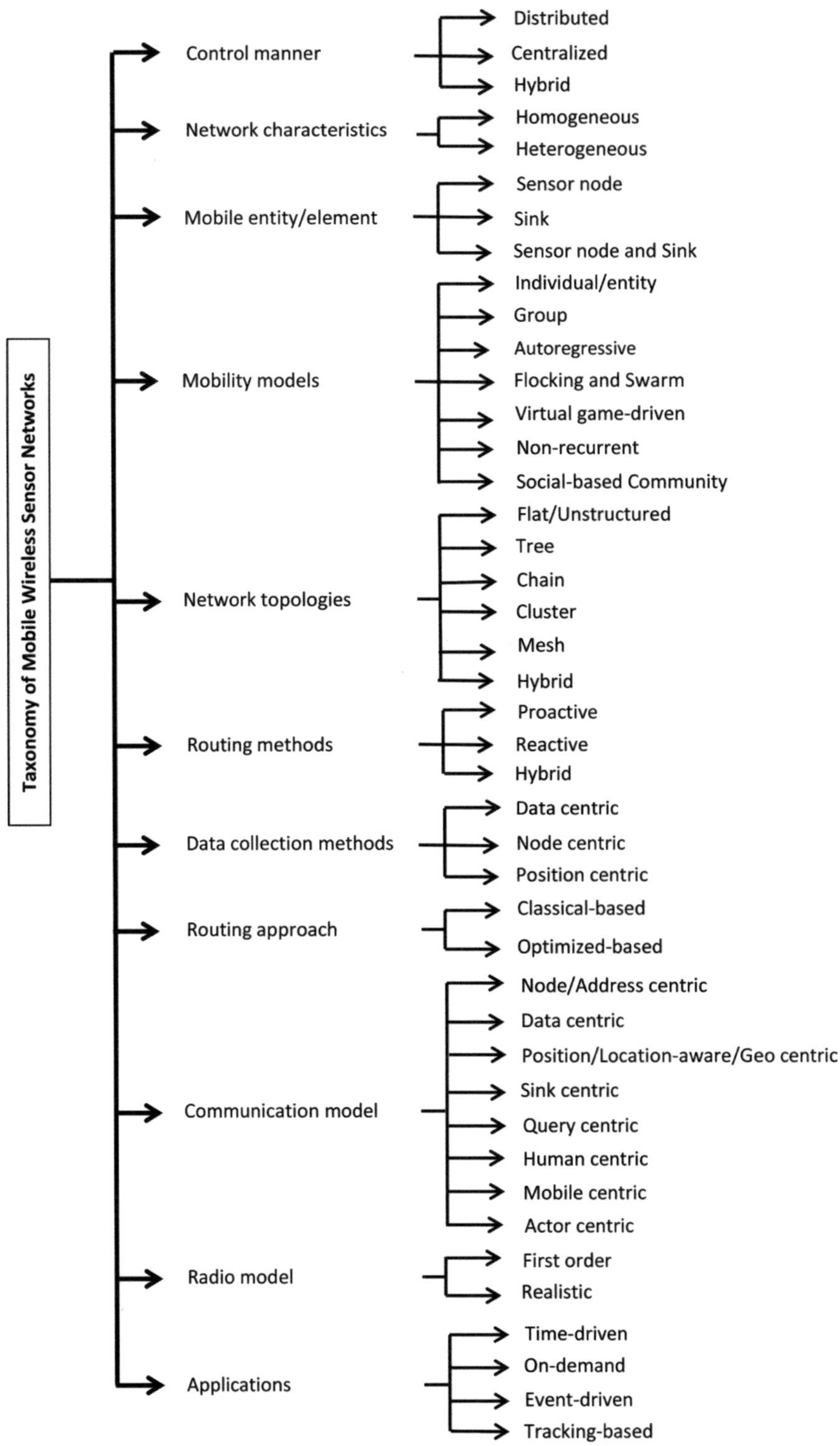

Figure 3. Taxonomy of mobile wireless sensor networks.

7.1.1. T-LEACH

Like LEACH, T-LEACH protocol [28] establishes a hierarchical topology for the large scale, dynamic and uneven distributed mobile WSNs. It uses the tree topology; power consumption mechanism and multi-hop transmit scheme to balance the whole network energy consumption as well as to increase the packet delivery rate. T-LEACH is implemented in two phases such as the topology construction stage and topology maintenance stage. On first, the topology construction stage establishes a data aggregation tree, cluster structure, and multi-hop transmit mechanism. Secondly, the topology maintenance stage follows the multi-hop transmits scheme, member nodes mobile reaction mechanism and cluster mobile reaction mechanism to establishes the stable network. From the simulation results, the T-LEACH protocol can effectively establish and maintain the topology structure of the dynamic and uneven distributed large-scale MWSNs in terms of packet delivery ratio and average energy consumption when compared to LEACH and cluster based routing (CBR) mobile LEACH.

7.1.2. Mobile LEACH/LEACH mobile

Mobile LEACH [29] is a mobility-centric protocol which is designed for mobile wireless sensor networks. Mobile LEACH operations are very similar to the static LEACH operations. But, the mobile LEACH allows the inclusion of the mobile sensor nodes with non-cluster heads on a setup phase and also rearrange the cluster of minimum energy consumption. After the cluster formation, the cluster head assigns a time slot for all the sensor nodes in its cluster. Also, the cluster members turn off the radio except during its transmit time to minimize the energy dissipation of the individual sensors. From the experimental results, Mobile LEACH outperforms LEACH by reducing the data packet loss for the mobile nodes. But, Mobile LEACH has the trade-off that it increases the unwanted energy dissipations while compared to LEACH.

Mezghani and Abdellaoui [30] proposed mobile mono-hop LEACH and mobile multi-hop LEACH for improving the lifetime of the mobile LEACH for the MWSNs. In mobile mono-hop LEACH, each mobile sensor node may directly communicate with the sink which is suitable for the small-scale indoor environment. Then, the mobile multi-hop LEACH is designed to support for some large scale outdoor applications. Simulation results proved that the proposed protocols (mobile and static LEACH for the mono and multi-hop architecture) improve the performance of the heterogeneous MWSNs as in terms of network lifetime, exchanged packet rate, delay and loss packet rate [31].

7.1.3. LEACH-mobile-enhanced (LEACH-ME)

LEACH purely considers the sensor nodes residual energy level for selecting the cluster head on each round, therefore, LEACH is not suitable for dynamic networks. But, LEACH-ME elects the cluster head based on the sensor node mobility and energy level [32]. Also, LEACH-ME maintains some information about the sensor nodes such as role of the node, mobility factor, cluster members list and TDMA schedule. Even the sensor node maintains all these four information, the mobility factor is a prime key to select the cluster head. Let, the mobility factor is calculated based on the transition count and the concept of remoteness. Because of these pieces of information, each cluster head can form a group of cluster members of minimum node mobility. Also, LEACH-ME ensures that the clusters are distributed minimally while the cluster heads on mobility. Simulation results show that LEACH-ME performs well than mobile LEACH in terms of average successful communication, normalized performance, computational overhead, and energy overhead with respect to mobility factor.

7.2. Mobile sink-based routing protocol (MSRP)

MSRP [33] is proposed to prolong the network life time of the cluster based wireless sensor networks. Usually, the sensor nodes very close to the sink should forward a large number of data packets when compared to far away sensor nodes, and they may drain their residual energy very quickly. This problem is termed as hotspot problem. To avoid such kind of issues, the sensing region is the portion of clusters and that prolongs the lifetime of the network. In MSRP, the mobile sink is considered instead of static sink, and that visits each cluster to collect the sensed data onto their cluster head. Now, the mobile sink collects the residual energy information about the cluster heads and that moves to the higher energy cluster heads.

The MSRP protocol operations consist of the setup phase and the steady state phase. In the setup phase, the whole network is portioned into clusters and the mobile sink advertises its location to cluster heads by broadcasting the beacon message for the registration process. Further, it is divided into three sub-phases such as initialization, mobile sink advertisement, and cluster head registration. After completion of the setup phase, the steady state phase is initiated. In the steady state phase, sink collects the data onto the registered cluster head and the cluster head collects the date from the cluster members. Then, it can be divided into three sub-phases such as TDMA scheduling, forwarding to sink and sink movement. Based on the residual energy of the cluster head, the mobile sink frequently visits all cluster heads in a network and that collects the data from among them. Simulation results proved that the MSRP reduces the energy consumption among the mobile sensor nodes and solves the hot spot problem due to changing the one-hop neighbor nodes of the mobile sink.

7.3. Mobility adaptive cross-layer routing (MACRO)

The MACRO [34] is developed to meet some essential requirements of the MWSNs such as end-to-end reliability, minimal power consumption, and packet delay. A single layer protocol development cannot provide some optimal solutions to the large scale MWSNs. Therefore, the authors developed a cross-layer interaction MACRO protocol, which combines five open systems interconnection (OSI) reference layers features such as application, transport, network, media access control (MAC) and physical layers into the all-in-one protocol. The MACRO protocol design consists of route discovery, data forwarding, and route management algorithms, which provides reliable quality links to frequent topology changes. Also, it reduces the node failures, unwanted control packets flooding and serious congestion of the MWSNs. However, the route discovery process of the large-scale mobile sensor networks may cause more delay due to the higher number of mobile sensor nodes as well as frequent topology changes. The simulation results prove that MACRO performs well rather than the classical CBR-mobile and LEACH-mobile in the aspects of packet delivery ratio (PDR) and end-to-end packet delay.

7.4. Energy management algorithm in a WSN with multiple sinks (EMMS)

An EMMS [35] is proposed to improve the residual energy utilization and the quality of data transmission of the MWSNs. EMMS protocol operation for the energy management MWSN with a multiple mobile sink is considered into two stages. (i) Find a closed tour of each mobile sink. i.e., a length of the closed tour of each mobile sink is roughly equal. (ii) Determine the sojourn locations of each mobile sink on the found closed tour. Then, build a routing tree rooted at each sojourn location as well as the sojourn time of the location for the mobile sink.

In EMMS, there are two types wireless transmission interfaces are used on each mobile sink: (i) low power wireless interface—communicate with sensor nodes within a sensor network; (ii) high bandwidth wireless interface—communicate with a third-party network for remote data transfer purpose. Like LEACH, EMMS energy management operations are considered into rounds. Each round consists of a routing tree construction, a sojourn time calculations at each sojourn location, and a data collection and sensing data transmission. Based on the residual energy of the sensor nodes, routing tree construction phase initiates a routing tree at each sojourn location on the closed tour of the mobile sink. Then, the sojourn time-calculation starts to calculate the members of the constructed tree and that estimates a sojourn time at each sojourn location on the closed tour of each sink. In a data collection and sensing data transmission phase, the sensor nodes forward their sensed data onto the mobile sink via the constructed routing tree and that transmits the sensed data to the remote monitoring center. At the end of the data transmission stage, the mobile sink collects the residual energy information about the sensor nodes in the neighbor set and then travels through its next sojourn location. The simulation results proved that the EMMS improves the residual energy usage as well as the data transmission quality of the MWSN significantly.

7.5. Artificial bee colony (ABC) based data collection for large-scale MWSNs

Zhang et al. [36] proposed the ABC algorithm, which considers three "bee" (i.e., onlookers, scouts, and employed bees) groups in the "colony". The ABC algorithm represents the population of bees to identify the optimal path and each bee represents a position in the search space. A bee always waits on the "dance" area to pick a honey source in an onlooker, randomly searches a scout and return to the previously visited honey source is represented as an employed bee. Let, the position of the honey sources signifies the possible solution to the optimization problem. In addition, the amount of "nector" of a honey source relates to the quality (i.e., fitness) of the correlated problem. Furthermore, the first half of the ABC algorithm denotes the employed bees and the second half signifies the onlooker bees. The ABC algorithm operation is divided into four main steps: initialization, population updating, bee source selection and population elimination.

Yue et al. [37] proposed an optimization based ABC mechanism, which uses the mobile sink to identify the optimal moving path as well as the routing path to collect the data onto the mobile sensor nodes of the large-scale MWSNs. Also, the ABC algorithm is designed to investigate the data latency of the mobile sink on three aspects: data collection maximization, mobile path length minimization, and network reliability optimization. Like LEACH, the ABC algorithm operation consists of an initial phase and a data collection phase on each round. In the initial phase, the mobile sink uses the network topology information to identify the optimal clusters of the best cluster head nodes as well as to establish a routing tree among the cluster head nodes. After the completion of the initial phase, the data collection is initiated. Here, each cluster head collects the sensed data from their cluster members. Then, the mobile sink responds immediately to the cluster head, which collects the data from the cluster head.

Further, the mobile sink identifies and picks the next cluster heads position according to the current network environment parameters along with the mobility, and that helps to identify the shortest path to each cluster head. The simulation results show that ABC is compared with the random walk and ant colony algorithms, the ABC algorithm can effectively improve certain metrics of the MWSNs such as data collection efficiency, residual energy utilization, reliability, and that prolongs the lifetime of the MWSNs.

7.6. Mobility-based clustering (MBC) protocol

Deng et al. [38] developed a mobility-based clustering (MBC) Protocol to improve the performance of the MWSNs. Like LEACH, the MBC protocol operation consists of a setup phase and a steady-state phase on each round. In the setup phase, all the sensor nodes have an equal chance to elect the cluster head based on the threshold value (i.e., residual energy and mobility). Also, MBC considers the connection time for cluster formation process, which builds a more reliable path based on the stability or an availability of each link between the cluster members and the cluster head.

7.7. Cluster independent data collection tree protocol

Velmani and Kaarthick developed a cluster independent data collection tree (CIDT) scheme [39], to provide reliable guaranteed end-to-end communication for large-scale MWSNs. CIDT is a unique methodology of the hybrid logical scheme, which utilizes to intra-cluster communication and data collection tree (DCT) communication for cluster and tree topologies respectively. The protocol design helps to improve the QoS parameters in terms of data collection, energy consumption, delay, PDR, throughput and network lifetime for large scale mobile WSNs. In CIDT, each sensor node picks the cluster head with better connection time, then the cluster head collects the data packets of the cluster members in an allocated time slot. After the cluster head election, base station initiates the DCT to elect one-hop neighbor DCN or current DCN which picks one-hop neighbor DCN or another cluster head with good coverage distance, maximum connection time and minimized network traffic.

The protocol operation consists of set-up phase and steady-state phase. During the set-up phase, a sensor elects itself as a cluster head among them based on its threshold value (i.e., flag, residual energy, and mobility). Furthermore, the cluster member joins with the one-hop cluster head during cluster formation based on the estimated connection time, received signal strength (RSS) and robustness in the connection.

Subsequently, the DCT communication is initiated by the base station to construct a data collection tree, which selects the data collection node (DCN) for covering the entire cluster heads. Here, the DCN does not participate in sensing for this particular round, which simply collects the sensed data and aggregates the data packets belonging to one-hop cluster head and the DCN, thereby forwarding the data packet to the base station through DCT. In order to keep the lifetime of the entire network to be well-balanced, new cluster heads and DCNs are elected for every round. Even when the sensor nodes are on high mobility, the DCN keeps communication with the cluster head, and CIDT no needs to update the tree structure for that particular round. Moreover, CIDT reduces energy consumption, link failure, end-to-end delay and traffic of cluster head due to forwarding the data with DCT. Minor complexity is involved in a sink to create a tree structure which reduces the energy consumption of cluster head. Also, CIDT helps to minimize the frequent cluster formation and maintain a cluster for a considerable amount of time.

In steady state phase, each cluster member sends its gathered data to the cluster head during its allocated time slot, and the cluster head aggregates the data and forwards it to the base station via intermediate cluster heads by direct sequence spread spectrum (DSSS) technique. Simulation results prove that CIDT achieves outstanding performance when compared to LEACH, HEED, and MBC in terms of PDR, throughput, delay and total energy consumption even in mobile sensor ambiance.

7.8. Velocity energy-efficient and link-aware cluster-tree (VELCT)

VELCT [40] is a hybrid topology (i.e., cluster and tree topology) based energy efficient routing scheme, which has been particularly designed to improve the network performance, data collection and lifetime of large scale mobile WSNs. VELCT is an enhancement of CIDT technique, which effectively mitigates the existing issues in network topologies such as residual energy consumption, coverage problem, critical node occurrence, RSS, traffic, connection time, tree intensity, scalability, fault tolerance, delay, throughput, PDR, mobility and network lifetime. The proposed VELCT scheme constructs the data collection tree (DCT). In DCT, a few sensor nodes in a network have been assigned as the data collection node (DCN) based on the position of the cluster head and that does not participate in sensing on this particular round. Here, the purpose of the DCN consideration is to be collecting the data packets of the cluster head and delivers it to the sink. The VELCT protocol minimizes residual energy consumption and reduces the end-to-end delay and traffic in cluster head in large scale MWSNs due to the efficient usage of the DCT. The major strengths of the VELCT algorithm are: reduce the residual energy consumption of the cluster head, constructs a simple tree structure, avoiding frequent cluster formation and maintains the cluster of a considerable amount of time.

The overall operation of the proposed protocol is partitioned into set-up phase and steady-state phase. The set-up phase consists of intra-cluster and DCT communication, which estimates the node position, threshold value, RSSI, connection time and robustness of connection. Moreover, DCT estimates the traffic, tree intensity, delay, mobility and throughput for every round. The DCN of the DCT does not participate in sensing for this particular round, but simply collects the data packets of the cluster head or adjacent DCNs, further to deliver the data packet of the sink through DCT, thereby the DCN acts as an ordinary sensor node from next successive rounds. The steady state phase is similar to that of CIDT.

The strength of VELCT protocol is, constructing a simple tree structure of cluster heads that maintain the cluster of a considerable amount of time, which reduces the energy consumption and control packet overhead, thereby avoid the bottleneck problem with the cluster head level and frequent clustering on mobility ambiance. Simulation results demonstrate that VELCT could yield better performance in terms of energy consumption, throughput, delay and PDR with reduced network traffic when compared to energy-efficient data collection protocol based on tree (EEDCP-TB), chain oriented sensor network (CREEC), cluster-tree data gathering algorithm (CTDGA), MBC and CIDT even in high mobility ambiance.

8. Conclusions

This chapter presented in detail the research carried on the large-scale mobile wireless sensor networks. Also, analyzed the major technical challenges as well as the research issues in designing hardware architectures, mobility, algorithms, and protocols for the large-scale MWSNs. We discussed most of the existing literature works of the MWSNs such as mobile sensor node architecture, mobility, topology, and routing protocols. Finally, we classified the existing mobility models and the taxonomy of MWSNs.

Author details

Velmani Ramasamy

Address all correspondence to: velmanir@gmail.com

Department of Electrical and Computer Engineering, Woldia University, Woldia, Ethiopia

References

[1] Karl H, Willig A. Protocols and Architectures for Wireless Sensor Networks. New York: Wiley; 2005 526 p

[2] Akyildiz IF, Su W, Sankarasubramaniam Y, Cayirci E. A survey on sensor networks. IEEE Communications Magazine. 2002;**40**(8):102-114

[3] Chelouah L, Semchedine F, Bouallouche-Medjkoune L. Localization protocols for mobile wireless sensor networks: A survey. Computers and Electrical Engineering. 2017;1-19. DOI: 10.1016/j.compeleceng.2017.03.024

[4] Silva R, Zinonos Z, Silva JS, Vassiliou V. Mobility in WSNs for critical applications. In: Proceedings of the 16th IEEE Symposium on Computers and Communications (ISCC '11); Corfu. Greece; 2011. pp. 451-456

[5] Mamun Q. A qualitative comparison of different logical topologies for wireless sensor networks. Sensors. 2012;**12**(11):4887-14913. DOI: 10.3390/s121114887

[6] Velmani R, Kaarthick B. Novel topology management schemes for effective energy utilization in large scale mobile wireless sensor networks [Thesis]. India: Shodhganga; 2016 179 p. Available from: http://shodhganga.inflibnet.ac.in/handle/10603/142867

[7] Rezazadeh J, Moradi M, Ismail AS. Mobile wireless sensor networks overview. International Journal of Computer Communications and Networks. 2012;**2**(1):17-22

[8] Wu C-H, Chung Y-C. Heterogeneous wireless sensor network deployment and topology control based on irregular sensor model. In: Cérin C, Li KC, editors. Second International Conference on Grid and Pervasive Computing. GPC 2007. Lecture Notes in Computer Science.; May 2-4; Paris, France. Berlin, Heidelberg: Springer; 2007. pp. 78-88. DOI: 10.1007/978-3-540-72360-8_7.

[9] Sankar S, Ranganathan H, Venkatasubramanian S. A study on next generations heterogeneous sensor networks. In: 2009 5th IEEE GCC Conference & Exhibition; 17-19 March 2009. Kuwait City, Kuwait: IEEE; 2009. pp. 1-4. DOI: 10.1109/IEEEGCC.2009.5734243.

[10] Wang J, Yang X, Zhang Z, Li B, Kim J-U. A survey about routing protocols with mobile sink for wireless sensor network. International Journal of Future Generation Communication and Networking. 2014;**7**(5):221-228

[11] Romer K, Mattern F. The design space of wireless sensor networks. IEEE Wireless Communications. 2004;**11**(6):54-61. DOI: 10.1109/MWC.2004.1368897

[12] Gungor VC, Hancke GP. Industrial wireless sensor networks: Challenges, design principles, and technical approaches. IEEE Transactions on Industrial Electronics. 2009;**56**(10):4258-4265. DOI: 10.1109/TIE.2009.2015754

[13] Srivastava N. Challenges of next-generation wireless sensor networks and its impact on society. Journal of Telecommunications. 2010;**1**(1):128-133

[14] Rathee A, Singh R, Nandini A. Wireless sensor network—challenges and possibilities. International Journal of Computer Applications. 2016;**140**(2):

[15] Sabor N, Sasaki S, Abo-Zahhad M, Ahmed SM. A comprehensive survey on hierarchical-based routing protocols for mobile wireless sensor networks: Review, taxonomy, and future directions. Wireless Communications and Mobile Computing. 2017;**2017**:23, Article ID 2818542. DOI: 10.1155/2017/2818542

[16] Camp T, Boleng J, Davies V. A survey of mobility models for ad hoc network research. Wireless Communications and Mobile Computing. 2002;**2**(5):483-502. DOI: 10.1002/wcm.72

[17] Vasanthi V, Romen Kumar M, Ajith Singh A, Hemalatha M. A detailed study of mobility models in wireless. Journal of Theoretical and Applied Information Technology. 2011;**33**(1):7-14

[18] Roy RR. Handbook of Mobile Ad Hoc Networks for Mobility Models. US: Springer; 2011. 1103 p. DOI: 10.1007/978-1-4419-6050-4

[19] Niculescu D. Communication paradigms for sensor networks. IEEE Communications Magazine. 2005;**43**(3):116-122. DOI: 10.1109/MCOM.2005.1404605

[20] Lai X, Liu Q, Wei X, Wang W, Zhou G, Han G. A survey of body sensor networks. Sensors. 2013;**13**(5):5406-5447

[21] Sara GS, Sridharan D. Routing in mobile wireless sensor network: A survey. Telecommunication Systems. 2014;**57**(1):51-79

[22] Yu S, Zhang B, Li C, Mouftah HT. Routing protocols for wireless sensor networks with mobile sinks: A survey. IEEE Communications Magazine. 2014;**52**(7):150-157

[23] Fuller R, Koutsoukos XD. Mobile entity localization and tracking in GPS-less environments. In: Second International Workshop, MELT 2009, A Survey on Localization for Mobile Wireless Sensor Networks; September 30 2009; Orlando, FL, USA. LNCS; 2009. pp. 235-254

[24] Liu X. A survey on clustering routing protocols in wireless sensor networks. Sensors. 2012;**12**(8):11113-11153

[25] Kim D-S, Chung Y-J. Self-organization routing protocol supporting mobile nodes for wireless sensor network. In: Proceedings of the First International Multi-Symposiums on Computer and Computational Sciences (IMSCCS '06); 20-24 April 2006, Washington DC. USA: IEEE Computer Society; 2006. pp. 622-626. DOI: 10.1109/IMSCCS.2006.265

[26] Awwad SAB, Ng CK, Noordin NK, Rasid MF. Cluster based routing protocol for mobile nodes in wireless sensor network. In: Proceeding of the International Symposium on Collaborative Technologies and Systems, 2009 (CTS '09);18-22 May 2009. Baltimore, MD, USA: IEEE; 2009. DOI: 10.1109/CTS.2009.5067486

[27] Nayebi A, Sarbazi-Azad H. Performance modeling of the LEACH protocol for mobile wireless sensor networks. Journal of Parallel and Distributed Computing. 2011;**71**(6):812-821. DOI: 10.1016/j.jpdc.2011.02.004

[28] Qi Z, Mini Y. A routing protocol for mobile sensor network based on LEACH. In: Proceedings of the 10th International Conference on Wireless Communications, Networking and Mobile Computing (WiCOM 2014); 26-28 September 2014; Beijing. China: IET; 2015. p. 473-477. DOI: 10.1049/ic.2014.0148

[29] Gu Y, Zhao L, Jing D. A novel routing protocol for mobile nodes in WSN. In: Proceeding of the 2012 International Conference on Control Engineering and Communication Technology (ICCECT); 7-9 December 2012; Liaoning. China: IEEE; 2013. pp. 624-627. DOI: 10.1109/ICCECT.2012.17

[30] Mezghani O, Abdellaoui M. Improving network lifetime with mobile LEACH protocol for wireless sensors network. In: Proceedings of the 15th International Conference on Sciences and Techniques of Automatic Control & Computer Engineering-STA'2014; December 21-23; Hammamet. Tunisia: IEEE; 2014. pp. 613-619

[31] Souid I, Chikha HB, Monser ME, Attia R. Improved algorithm for mobile large scale sensor networks based on LEACH protocol. In: Proceeding of the 22nd International Conference on Software, Telecommunications and Computer Networks (SoftCOM), 2014;17-19 September 2014; Split. Croatia: IEEE; 2015. DOI: 10.1109/SOFTCOM.2014.7039097

[32] Santhosh Kumar G, Vinu PMV, Jacob KP. Mobility metric based LEACH-mobile protocol. In: Proceeding of the 16th International Conference on Advanced Computing and Communications, 2008 (ADCOM 2008); 14-17 December 2008; Chennai. India: IEEE; 2009. pp. 248-253. DOI: 10.1109/ADCOM.2008.4760456

[33] Nazir B, Hasbullah H. Mobile sink based routing protocol (MSRP) for prolonging network lifetime in clustered wireless sensor network. In: Proceeding of the 2010 International Conference on Computer Applications and Industrial Electronics (ICCAIE 2010); 5-7 December 2010; Kuala Lumpur. Malaysia: IEEE; 2010. pp. 624-629. DOI: 10.1109/ICCAIE.2010.5735010

[34] Cakici S, Erturk I, Atmaca S, Karahan A. Wireless personal communications. A Novel Cross-layer Routing Protocol for Increasing Packet Transfer Reliability in Mobile Sensor Networks. 2014;**77**(3):2235-2254. DOI: 10.1007/s11277-014-1635-0

[35] Shi J, Wei X, Zhu W. An efficient algorithm for energy Management in Wireless Sensor Networks via employing multiple mobile sinks. International Journal of Distributed Sensor Networks. 2016;**2016**:9 p. Article ID 3179587. DOI: 10.1155/2016/3179587

[36] Zhang X, Zang X, Yuen SY, Ho SL, Fu WN. An improved artificial bee colony algorithm for optimal design of electromagnetic devices. IEEE Transactions on Magnetics. 2013;**49**(8):811-4816

[37] Yue Y, Li J, Fan H, Qin Q. Optimization-based artificial bee Colony algorithm for data collection in large-scale mobile wireless sensor networks. Journal of Sensors. 2016;**2016**:12 p Article ID 7057490. DOI: 10.1155/2016/7057490

[38] Deng S, Li J, Shen L. Mobility-based clustering protocol for wireless sensor networks with mobile nodes. IET Wireless Sensor Systems. 2011;**1**(1):39-47. DOI: 10.1049/iet-wss.2010.0084

[39] Velmani R, Kaarthick B. An efficient cluster-tree based data collection scheme for large mobile wireless sensor networks. IEEE Sensors Journal. 2015;**15**(4):2377-2390. DOI: 10.1109/JSEN.2014.2377200

[40] Velmani R, Kaarthick B. An energy efficient data gathering in dense mobile wireless sensor networks. ISRN Sensor Networks. 2014;**2014**:10 p. Article ID 518268. DOI: 10.1155/2014/518268

Chapter 2

Routing Protocols for Wireless Sensor Networks (WSNs)

Noman Shabbir and Syed Rizwan Hassan

Additional information is available at the end of the chapter

http://dx.doi.org/10.5772/intechopen.70208

Abstract

Wireless sensor networks (WSNs) are achieving importance with the passage of time. Out of massive usage of wireless sensor networks, few applications demand quick data transfer including minimum possible interruption. Several applications give importance to throughput and they have not much to do with delay. It all rest on the applications desires that which parameter is more favourite. The knowledge of network structure and routing protocol is very important and it should be appropriate for the requirement of the usage. In the end a performance analysis of different routing protocols is made using a WLAN and a ZigBee based Wireless Sensor Network.

Keywords: routing protocols, WSN, DSR, AODV, OLSR, WLAN, ZigBee

1. Introduction

The routing protocol is a process to select suitable path for the data to travel from source to destination. The process encounters several difficulties while selecting the route, which depends upon, type of network, channel characteristics and the performance metrics.

The data sensed by the sensor nodes in a wireless sensor network (WSN) is typically forwarded to the base station that connects the sensor network with the other networks (may be internet) where the data is collected, analyzed and some action is taken accordingly.

In very small sensor networks where the base station and motes (sensor nodes) so close that they can communicate directly with each other than this is single-hop communication but in most WSN application the coverage area is so large that requires thousands of nodes to be placed and this scenario requires multi-hop communication because most of the sensor nodes are so far from the sink node (gateway) so that they cannot communicate directly with the base station. The single-hop communication is also called direct communication and multi-hop communication is called indirect communication.

In multi-hop communication the sensor nodes not only produce and deliver their material but also serve as a path for other sensor nodes towards the base station. The process of finding suitable path from source node to destination node is called routing and this is the primary responsibility of the network layer.

2. Routing challenges in WSNs

The design task of routing protocols for WSN is quite challenging because of multiple characteristics, which differentiate them, from wireless infrastructure-less networks. Several types of routing challenges involved in wireless sensor networks. Some of important challenges are mentioned below:

- It is almost difficult to allocate a universal identifiers scheme for a big quantity of sensor nodes. So, wireless sensor motes are not proficient of using classical IP-based protocols.
- The flow of detected data is compulsory from a number of sources to a specific base station. But this is not occurred in typical communication networks.
- The created data traffic has significant redundancy in most of cases. Because many sensing nodes can generate same data while sensing. So, it is essential to exploit such redundancy by the routing protocols and utilize the available bandwidth and energy as efficiently as possible.
- Moreover wireless motes are firmly restricted in relations of transmission energy, bandwidth, capacity and storage and on-board energy. Due to such dissimilarities, a number of new routing protocols have been projected in order to cope up with these routing challenges in wireless sensor networks.

3. Design challenges in WSNs

There are some major design challenges in wireless sensor networks due to lack of resources such as energy, bandwidth and storage of processing. While designing new routing protocols, the following essentials should be fulfilled by a network engineer.

3.1. Energy efficiency

Wireless sensor networks are mostly battery powered. Energy shortage is a major issue in these sensor networks especially in aggressive environments such as battlefield etc. The performance of sensor nodes is adversely affected when battery is fallen below a pre-defined battery threshold level. Energy presents a main challenge for designers while designing sensor networks. In wireless sensor network, there are millions of motes. Each node in this network has restricted energy resources due to partial amount of power. So, the routing protocol should be energy efficient [1].

3.2. Complexity

The complexity of a routing protocol may affect the performance of the entire wireless network. The reason behind is that we have inadequate hardware competences and we also face extreme energy limitations in wireless sensor networks.

3.3. Scalability

As sensors are becoming cheaper day by day, hundreds or even thousands of sensors can be installed in wireless sensor network easily. So, the routing protocol must support scalability of network. If further nodes are to be added in the network any time then routing protocol should not interrupt this.

3.4. Delay

Some applications require instant reaction or response without any substantial delay such as temperature sensor or alarm monitoring etc. So, the routing protocol should offer minimum delay. The time needed to transmit the sensed data is required to be as little as possible in above cited WSN applications.

3.5. Robustness

Wireless sensor networks are deployed in very crucial and loss environments frequently. Occasionally, a sensor node might be expire or leaving the wireless sensor network. Thus, the routing protocol should be capable to accept all sorts of environments including severe and loss environments. The functionality of the routing protocol should be fine also [2].

3.6. Data transmission and transmission models

There are four modes of data transmission depending on the applications in wireless sensor networks namely as query driven, event driven and continuous type and hybrid type. A node begins to transmit the data only when sink creates the query or an event occurs in query driven model and event driven model. The data is sent out periodically in continuous transmission mode. The performance of the routing protocol is a function of network size and transmission media. So, transmission media of good quality enhances the network performance directly [3].

3.7. Sensor location

Another major challenge that is faced by wireless sensor network designers is to correctly locate of the sensor nodes. Most routing protocols use some localization technique to obtain knowledge concerning their locations. Global positioning system (GPS) receivers are used in some scenario.

4. Classification of routing protocols

The routing protocols define how nodes will communicate with each other and how the information will be disseminated through the network. There are many ways to classify the routing protocols of WSN. The basic classification of routing protocols is illustrated in **Figure 1**.

4.1. Node centric

In node centric protocols the destination node is specified with some numeric identifiers and this is not expected type of communication in Wireless sensor networks. E.g. Low energy adaptive clustering hierarchy (LEACH).

4.1.1. Low energy adaptive clustering hierarchy (LEACH)

LEACH is a routing protocol that organizes the cluster such that the energy is equally divided in all the sensor nodes in the network. In LEACH protocol several clusters are produced of sensor nodes and one node defined as cluster head and act as routing node for all the other nodes in the cluster.

As in routing protocols the cluster head is selected before the whole communication starts and the communication fails if there is any problem occurs in the cluster head and there is much chances that the battery dies earlier as compare to the other nodes in cluster as the fix cluster head is working his duties of routing for the whole cluster.

LEACH protocol apply randomization and cluster head is selected from the group of nodes so this selection of cluster head from several nodes on temporary basis make this protocol more long lasting as battery of a single node is not burdened for long.

Sensor nodes elect themselves as cluster head with some probability criteria defined by the protocol and announce this to other nodes

4.2. Data-centric

In most of the wireless sensor networks, the sensed data or information is far more valuable than the actual node itself. Therefore data centric routing techniques the prime focus is on the

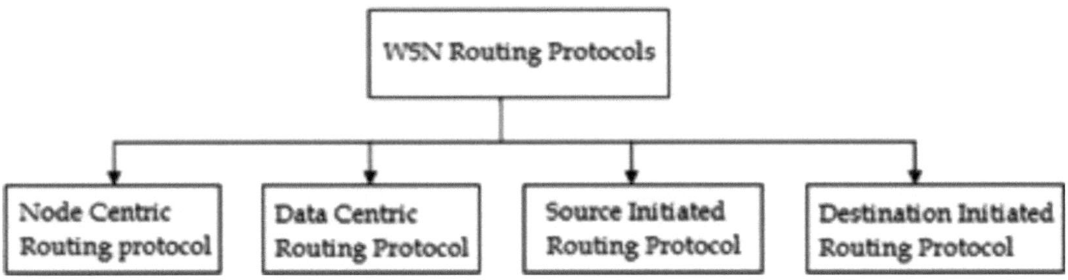

Figure 1. Basic classification of routing protocols.

transmission of information specified by certain attributes rather than collecting data from certain nodes.

In data centric routing the sink node queries to specific regions to collect data of some specific characteristics so naming scheme based on attributes is necessary to describe the characteristics of data. Examples are as follows:

4.2.1. Sensor protocols for information via negotiation (SPIN)

SPIN is abbreviation of sensor protocol for information via negotiation. This protocol is defined to use to remove the deficiency like flooding and gossiping that occurs in other protocols. The main idea is that the sharing of data, which is sensed by the node, might take more resources as compare to the meta-data, which is just a descriptor about the data sensed, by the node. The resource manager in each node monitors its resources and adapts their functionality accordingly.

Three messages namely ADV, REQ and DATA are used in SPIN. The node broadcast an ADV packet to all the other nodes that it has some data. This advertising node ADV message includes attributes of the data it has. The nodes having interests in data, which the advertising node has requested by sending REQ message, to the advertising node. On receiving the REQ message the advertising node send data to that node. This process continues when the node on reception of data generate an ADV message and send it. The whole model SPIN is shown in (**Figure 2**).

4.3. Destination-initiated (Dst-initiated)

Protocols are called destination initiated protocols when the path setup generation originates from the destination node. Examples are directed diffusion (DD) & LEACH.

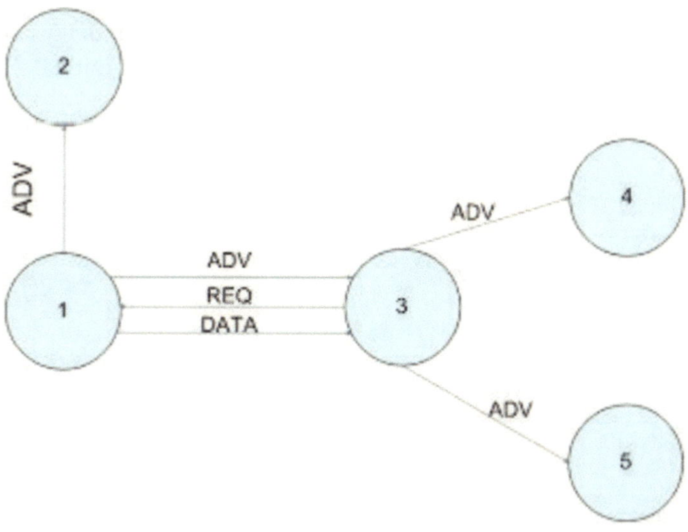

Figure 2. SPIN routing protocol.

4.3.1. Directed diffusion (DD)

Directed diffusion is a data centric routing technique. It uses this data centric technique for information gathering and circulating. This routing protocol is also energy efficient and energy saving protocol so that's why life time of the network is increased. All the communication in directed diffusion routing protocol is node to node so there is no need of addressing in this protocol.

4.4. Source-initiated (Src-initiated)

In these types of protocols the source node advertises when it has data to share and then the route is generated from the source side to the destination. Examples is SPIN.

5. Categories of routing protocols

In order to transmit data in sensor networks, there are two techniques being used. The one is referred to as Flooding and the other one is gossiping protocol. There is no need to use any routing algorithm and maintenance of topology. In the flooding protocol, upon reception of a data packet by sensor nodes, this data packet is broadcast to all other neighbors. The process of broadcasting is continued till any one of two following conditions is satisfied; the packet has reached successfully to its destination. And second condition is; maximum number of hops of a packet has reached [4].

The main advantages of flooding are ease of implementation and simplicity. The drawbacks are blindness of resources and overlapping and implosion. The gossiping protocol is somewhat advanced version of flooding protocol. In gossiping protocol, the sensor node, which is getting a data packet, transmits it to the arbitrarily selected neighbor. At the next turn, the sensing nodes again randomly pick another nodes and sends data to it. This process is continued again and again. The broadcasting is not used in gossiping protocol as it was used in flooding. In this way, implosion issue can be avoided easily. But delay is enhanced in this way. The main categories of the routing protocols are depicted in **Figure 3**.

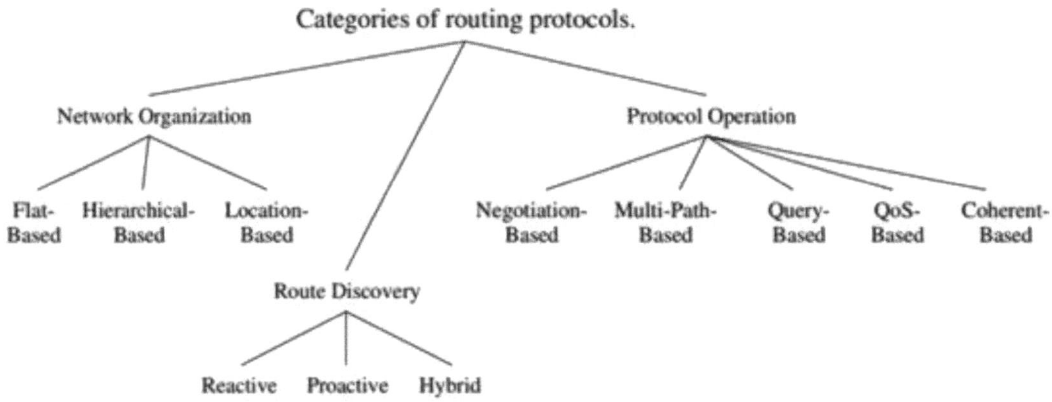

Figure 3. Categories of routing protocols.

5.1. Route discovery based routing protocols

Routing protocols are classified on the basis of process they used to discover the routes.

5.1.1. Reactive protocols

Reactive routing protocols do not maintain the whole network topology they are activated just on demand when any node wants to send data to any other node. So the routes are created on demand when queries are initiated. The most commonly used reactive routing protocols are as follows:

5.1.1.1. Ad-hoc on-demand distance vector routing system (AODV)

Ad-hoc on-demand distance vector (AODV) is reactive on request protocol. AODV is engineered for Mobile infrastructure-less networks. It employs the on-demand routing methodology for formations of route among network nodes. Path is established solitary when source node want to direct packs of data and pre-set route is maintained as long as the source node needs. That's why we call it as On-Demand. AODV satisfies unicast, multicast and broadcast routing. AODV routing protocol directs packets among mobile nodes of wireless ad-hoc network. AODV permits mobile nodes to pass data packets to necessary destination node via nodes of neighbor that are unable to connect link openly. The material of routing tables is switched intermittently among neighbor nodes and prepared for sudden updates [3].

AODV chooses shortest but round free path from routing table to transmit packets. Suppose if errors or variations come in nominated path, then AODV is intelligent enough to make a fresh new route for rest of communication.

5.1.1.2. Dynamic source routing (DSR)

Dynamic source routing (DSR) is a routing protocol used in wireless sensor networks developed at CMU in 1996. Dynamic source routing can be reactive or on demand. As its name shows that it uses source routing instead of routing tables. Routing in DSR is divided into two parts, route discovery and route maintenance.

Source node will initiate a route discovery phase and this phase consist of route request and route reply (RREP) messages. In DSR only destination node will reply with route reply RREP message to the source node unlike in AODV where every intermediate node would reply with route reply message RREP. And the purpose of next phase route maintenance is to avoid flooding of RREP messages and used for shortening of nodes between source and destination [6, 8].

5.1.2. Proactive protocols

They are also known as table driven routing protocols, because they maintains the routing tables for the complete network by passing the network information from node to node and the routes are pre-defined prior to their use and even when there is no traffic flow. The most commonly used algorithm is as follows:

5.1.2.1. Optimized link state routing (OLSR)

Optimized link state routing (OLSR) belongs to the category of proactive routing protocols and it uses table focused practice. The main drawback of OLSR is that it has a massive overhead. To compensate this delay, multipoint relays (MPRs) are used to overcome the large overhead. For data transmission, three adjutant nodes are used as MPRs by every node. No consistent control information is required as each node sends it alternatingly [6, 8].

5.1.3. Hybrid routing protocols

Hybrid Routing Protocols have the merits of proactive and reactive routing protocols by neglecting their demerits.

5.2. Network organization based routing protocols

Following protocols are based on the network organization of wireless sensor network.

5.2.1. Flat topology

Flat topology treats all nodes equally. Flat topology is mainly for homogeneous networks where all nodes are of same characteristics and have same functionality. Examples are:

- Gradient based routing (GBR)
- Cougar
- Constrained anisotropic diffusion routing (CADR)
- Rumor routing (RR)

5.2.2. Hierarchical based routing

Mostly heterogeneous networks apply hierarchical routing protocols where some nodes are more advance and powerful than the other nodes, but not always this is the case, sometimes in hierarchical (clustering) protocols sometimes the nodes are grouped together to form a cluster and the cluster head is assigned to every cluster, which after data aggregation from all the nodes, communicates with the base node .The clustering scheme is more energy efficient and more easily manageable. Examples are:

- Threshold sensitive energy efficient sensor network (TEEN)
- Adaptive threshold sensitive energy efficient sensor network (APTEEN)
- Low energy adaptive clustering hierarchy (LEACH)
- The power-efficient gathering in sensor information systems (PEGASIS)
- Virtual grid architecture routing (VGA)
- Self-organizing protocol (SOP)
- Geographic adaptive fidelity (GAF)

5.2.3. Location-based routing (geo-centric)

In location based routing the nodes have capability to locate their present location using various localization protocols. Location information helps in improving the routing procedure and also enables sensor networks to provide some extra services. Examples are:

- SPEED
- Geographical and energy aware routing (GEAR)
- SPAN

5.3. Operation based routing protocols

According to the operational basis the routing protocols are classified as:

- Multipath routing protocols
- Query based routing
- Negotiation based routing
- QoS-based routing
- Coherent routing

5.3.1. Multi-path routing protocol

Multi-path routing protocols provide multiple paths for data to reach the destination providing load balancing, low delay and improved network performance as a result. The multiple routing protocol also provide alternate path in case of failure of any path. Dense networks more interested in multiple path networks. To keep the paths alive some sort of periodic messages have to a send after some specific intervals hence multiple path routing is not more energy efficient. Multipath routing protocols are: [6]

- Multi path and Multi SPEED (MMSPEED)
- Sensor protocols for information via negotiation (SPIN)

5.3.2. Query based routing protocol

These type of routing protocols are mostly receiver-initiated. The sensor nodes will only send data in response to queries generated by the destination node. The destination node sends query of interest for receiving some information through the network and the target node sense the information and send back to the node that has initiated the request. The examples are [6]:

- Sensor protocols for information via negotiation (SPIN)
- Directed diffusion (DD)
- COUGAR

5.3.3. Negotiation based routing protocols

In these types of protocols to keep the redundant data transmission level at minimum, the sensor nodes negotiate with the other nodes a and share their information with the neighboring nodes about the resources available and data transmission decisions are made after the negotiation process. Examples are [6]:

- Sensor protocols for information via negotiation (SPAN)
- Sequential assignment routing (SAR)
- Directed diffusion (DD)

5.3.4. QoS based routing protocols

To get good Quality of Service these protocols are used. QoS aware protocols try to discover path from source to sink that satisfies the level of metrics related to good QoS like throughput, data delivery, energy and delay, but also making the optimum use of the network resources.

Examples are: [4, 6]

- Sequential assignment routing (SAR)
- SPEED
- Multi path and Multi SPEED (MMSPEED)

5.3.5. Coherent data processing routing protocol

In coherent data processing routing protocol the nodes perform minimum processing (time stamping, data compression etc.) on the data before transmitting it towards the other sensor nodes or aggregators. Aggregator performs aggregation of data from different nodes and then passes to the sink node.

5.4. Comparison of routing protocols of WSN

A detailed comparison of WSN routing protocols is given below in tabular form is shown in **Figure 4** [5].

5.5. Performance analysis of routing protocols

OPNET Modeler 14.5 network simulator is used to analyze AODV, DSR and OLSR routing protocols in WLAN based WSNs. These protocols are compatible in WLAN based WSNs and previous reseraches indicated that they have better performnace.Here, the perforrmance of these protocols will be evaluated in small, medium and large scale network against delay, throughput and network load. Small scale network contains 20 nodes, medium scale with 40 nodes and large scale network takes 80 nodes. The simulation model is represented in **Figure 5**. The general parameters for simulation scenarios are given in **Table 1**.

Simulation parameters	Values
No. of nodes	20, 40, 80
Simulation time	120 s
Simulation area	1000 m^2
Data rate of nodes	11 Mbps
Traffic	FTP (high load)
Routing protocols	AODV, DSR and OLSR

Table 1. Simulation parameters.

Routing Protocols	Classification	Power Usage	Data Aggregation	Scalability	Query Based	Overhead	Data delivery model	QoS
SPIN	Flat / Src-initiated / Data-centric	Ltd.	Yes	Ltd	Yes	Low	Event driven	No
DD	Flat/ Data-centric/ Dst-initiated	Ltd	Yes	Ltd	Yes	Low	Demand driven	No
RR	Flat	Low	Yes	Good	Yes	Low	Demand driven	No
GBR	Flat	Low	Yes	Ltd	Yes	Low	Hybrid	No
CADR	Flat	Ltd		Ltd	Yes	Low	Continuously	No
COUGAR	Flat	Ltd	Yes	Ltd	Yes	High	Query driven	No
ACQUIRE	Flat/ Data-centric	Low	Yes	Ltd	Yes	Low	Complex query	No
LEACH	Hierarchical / Dst-initiated /Node-centric	High	Yes	Good	No	High	Cluster-head	No
TEEN & APTEEN	Hierarchical	High	Yes	Good	No	High	Active threshold	No
PEGASIS	Hierarchical	Max	No	Good	No	Low	Chains based	No
VGA	Hierarchical	Low	Yes	Good	No	High	Good	No
SOP	Hierarchical	Low	No	Good	No	High	Continuously	No
GAF	Hierarchical / Location	Ltd	No	Good	No	Mod	Virtual grid	No
SPAN	Hierarchical / Location	Ltd	Yes	Ltd	No	High	Continuously	No
GEAR	Location	Ltd	No	Ltd	No	Mod	Demand driven	No
SAR	Data centric	High	Yes	Ltd	Yes	High	Continuously	Yes
SPEED	Location/Data centric	Low	No	Ltd	Yes	Less	Geographic	Yes

Figure 4. Comparison of routing protocols.

Now three network metrics are defined; End-to-End delay, throughput and network load. ETE delay is described by way of time engaged by an envelope to be communicated through a network from source to destination. It comprises retransmission delays on media access layer (MAC), packet transfer time and broadcast delay plus other delays

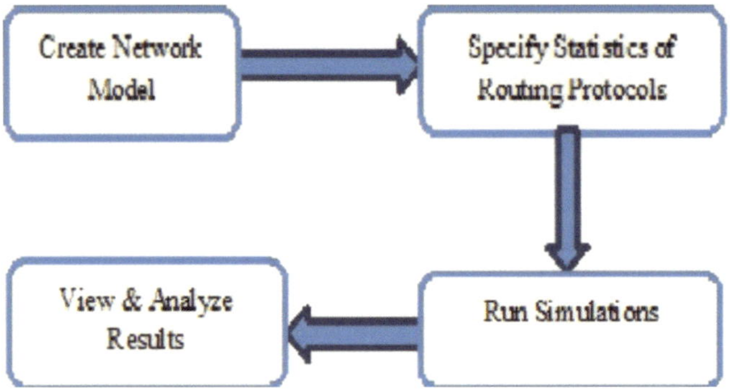

Figure 5. Simulation model.

at route discovery and conservation. The quantity of data transmission from source to destination network node in a given specified amount of time. It is dignified in byte per second. Network load (NL) shows net load, which indicates, in bits per second. Work load is sometimes also called as Network Congestion. When traffic load exceeds than link capacity then it is almost impossible for network to handle the traffic thus creating congestion in the network.

In simulations, there sensor networks are considered, firstly in a small scale network, 20 nodes are selected with one stationary WLAN server. These nodes are interconnected in star topology. Area of the network is 1000 × 1000 m. IPv4 scheme is applied to entirely nodes and File Transfer Protocol is used as great traffic load. Each WLAN node has data rate of 11 Mbps. Similarly, a medium scale network is with 40 nodes and large scale network is consisted of 80 nodes.

After running simulations, the following results are obtained. **Figures 6–8** depicts simulation results of delay, network load and throughput for AODV in small, medium and large scale networks, respectively. Delay is represented in seconds while throughtput and network load in bits per seconds.

The entire results of small, medium and large scale networks are mentioned below in **Table 2**. It is cocluded from the table that in terms of delay, the efficiency of OLSR is more than 100% in small and medium scale network as compared to the other two protocols while AODV is significantly (>50%) better in large networks. In case of network load, OLSR gives minimum load in all three scenarios. However, AODV gives best throughput in small scale network which is 40% more than DSR and 86% higher than OLSR. DSR is better than AODV and OLSR by a factor of 13 and 40% respectively, in medium scale network. Similarly, in large scale network it is better by a margin of 47 and 18%.

5.5.1. Performance analysis for a ZigBee based network

The same comparison can be made for a ZigBee based Wireless Sensor Network using AODV. ZIGBEE nodes use in lower data rates applications where we need a longer battery life. Through wireless sensor nodes provides higher data rates but their disadvantage is that they

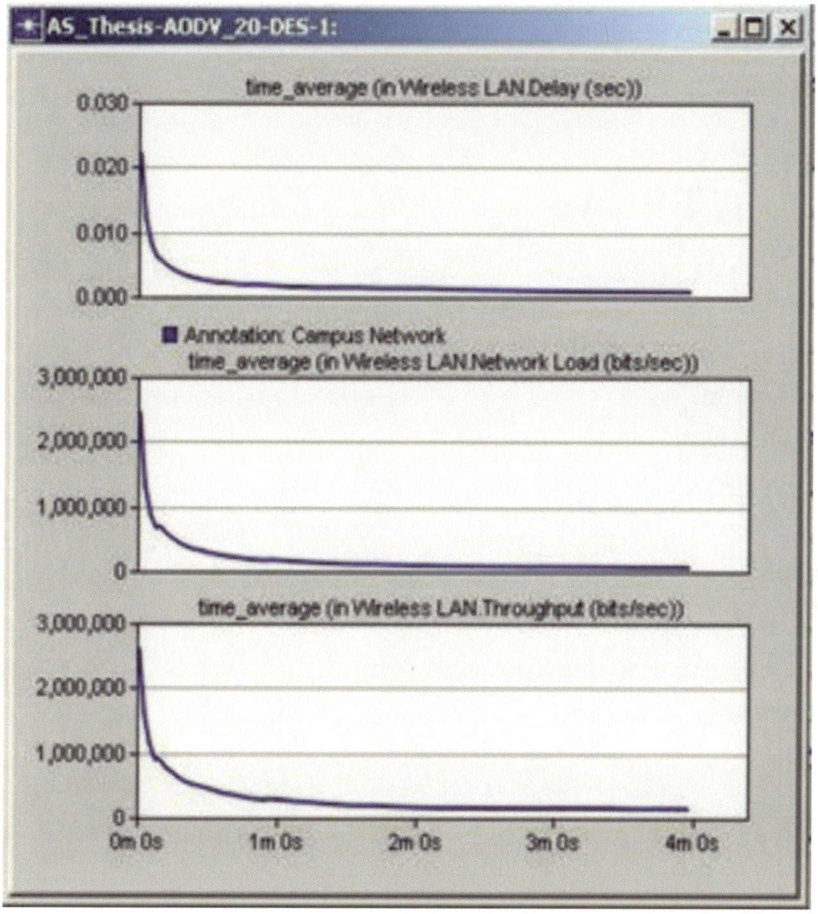

Figure 6. Simulation Results for AODV.

require higher power. So in those applications where we don't need higher data rates we use ZIGBEE because they increase the life of the network [7].

Figure 9 depicts that the end-to-end delay is higher in a network where we use ZIGBEE nodes. End-to-end delay starts from 0.060 s and then step up in the starting and then gets saturated at approximately 0.070 s. While in WSN nodes, End-to-end delay hardly increase from 0.010 s and throughput is lower in a ZIGBEE network as we can see in the **Figure 10**. From **Figure 10**, throughput increases linearly in the start and then gets stable at 6300 bits/s. So ZIGBEE nodes are used when there are concerns with the life span of network and economic issues because ZIGBEE is a low power, low cost devices.

5.6. Conclusion

Routing protocols plays a very significant part to produce interruption less and efficient communication between source and destination nodes. The performance, service and reliability of a network mostly depend on the selection of good routing protocol. Protocols being used

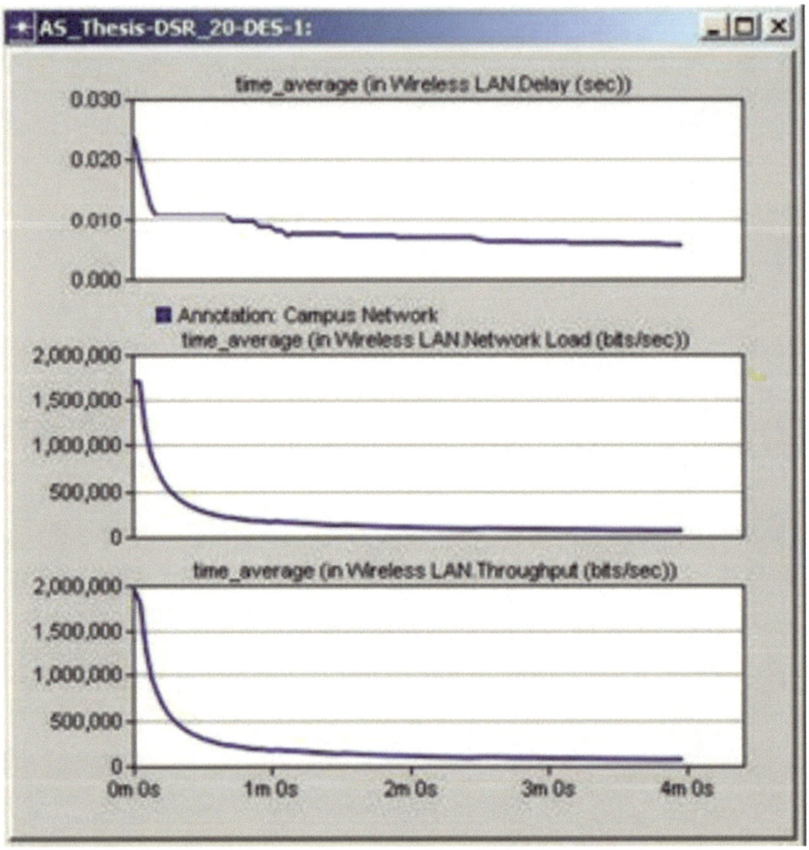

Figure 7. Simulation Results for DSR.

in Wireless sensor networks and ad hoc networks must be round-free. The routing protocols in WSN are classified in many different ways.

The categories of routing protocols are network based organization, operation and route discovery. Most of the applications of WSN uses route discovery base routing protocols e.g.

Nodes	Parameters	AODV	DSR	OLSR
20	Delay (s)	0.020	0.024	0.011
	Network load (Kbps)	2500	1700	1300
	Throughput (Kbps)	2800	2000	1500
40	Delay (s)	0.033	0.060	0.013
	Network load (Kbps)	3000	3000	2000
	Throughput (Kbps)	3700	4200	3000
80	Delay (s)	0.10	0.17	0.015
	Network load (Kbps)	3100	2900	2800
	Throughput (Kbps)	6200	13,000	11,000

Table 2. Simulation results.

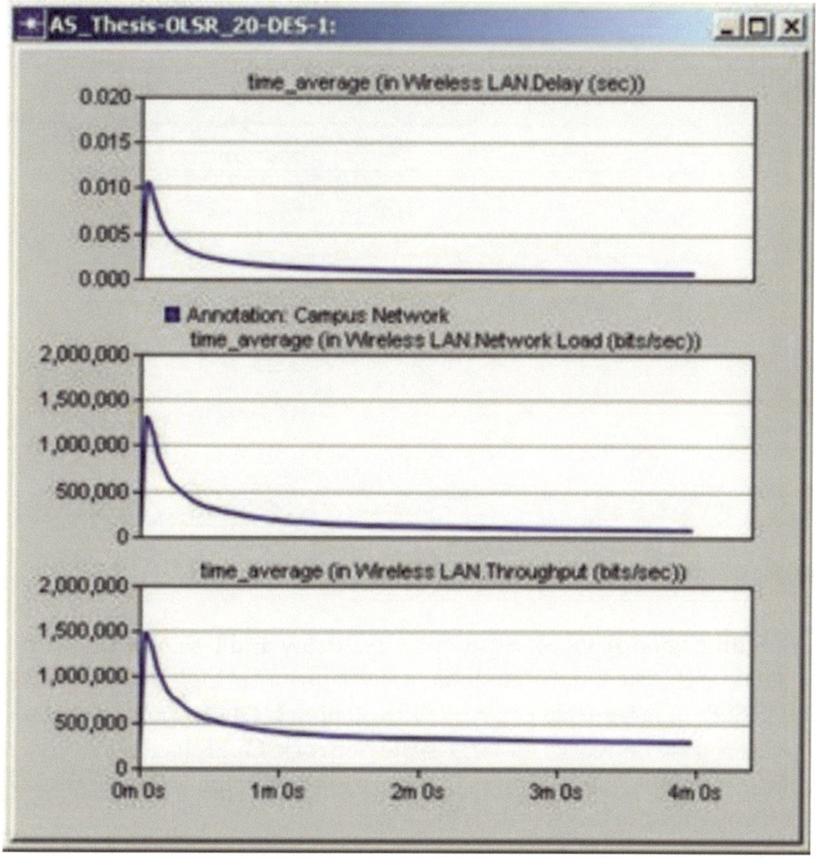

Figure 8. Simulation Results for a OLSR.

AODV, DSR & OLSR. The performance of these protocols is compared in different scenarios on the basis of throughput, delay and congestion.

In small scale network with 20 nodes, OLSR gives less jitter & less congestion/load as matched with AODV and DSR. AODV & DSR give high throughput than OLSR. In medium

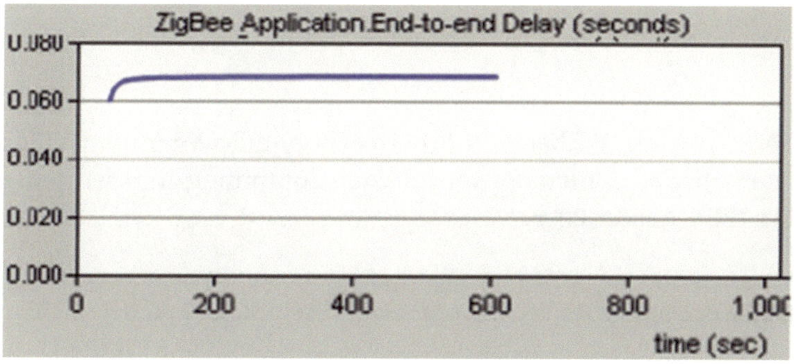

Figure 9. End-to-end delay in ZigBee.

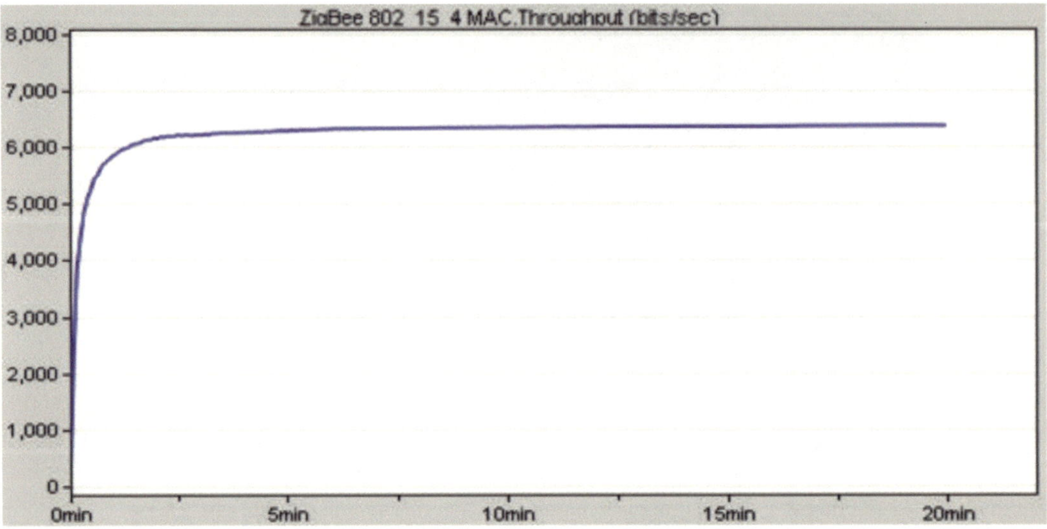

Figure 10. Throughput in ZigBee.

scale network with 40 nodes, OLSR again give less delay and less network load when compared with AODV and DSR. On the other hand, DSR provides high throughput as compared to AODV and OLSR. In large scale network with 80 nodes, OLSR shows same behavior as in small and medium scale networks. In large scale network, OLSR has less delay and network load than DSR and AODV. Interestingly, DSR give highest value for throughput. AODV has least value of throughput in large scale network.

Author details

Noman Shabbir* and Syed Rizwan Hassan

*Address all correspondence to: noman.shabbir@gcu.edu.pk

Department of Electrical Engineering, GC University Lahore, Pakistan

References

[1] G Yin, G Yang, W Yang, B Zhang, W Jin. An energy-efficient routing algorithm for wireless. In. International Conference on Internet Computing in Science and Engineering (ICICSE'08), IEEE, China; 2008.

[2] Zang Z, Qi JD, Cao YJ. A robust routing protocol in wireless sensor networks. In: IET International Conference on Wireless Sensor Network. China: IET; 2010. pp. 276-279

[3] Ehsan S, Hamdaoui B. A survey on energy-efficient routing techniques with QoS assurances for wireless multimedia sensor networks. IEEE Communications Surveys & Tutorials. 2012;**14**(2), PP. 265-278.

[4] Hassan SR. Performance analysis of ZigBee based wireless sensor networks [thesis]. Lahore: GCU; 2014. Available from: http://library.gcu.edu.pk/theses.htm

[5] Nawaz R. Performance analysis of WLAN based routing protocols [thesis]. Lahore: GCU; 2015. Available from: http://library.gcu.edu.pk/theses.htm

[6] Shabbir N, Nawaz R, Iqbal MN, Zafar J. Routing protocols for a small scale WLAN based wireless sensor networks. In: 9th International Conference on Sensing Technologies. New Zealand: IEEE; 2015

[7] Khan AA. A survey of routing protocol in wireless sensor networks [thesis]. Lahore: GCU; 2016. Available from: http://library.gcu.edu.pk/theses.htm

[8] Bakr BA, Lilien L. LEACH-SM: A protocol for extending wireless sensor network lifetime by management of spare nodes. In: International Symposium on Modelling and Optimization in Mobile, Ad Hoc and Wireless Networks (WiOpt); IEEE; New Jersey, Princeton; 2011

Chapter 3

Gradient Descent Localization in Wireless Sensor Networks

Nuha A.S. Alwan and Zahir M. Hussain

Additional information is available at the end of the chapter

http://dx.doi.org/10.5772/intechopen.69949

Abstract

Meaningful information sharing between the sensors of a wireless sensor network (WSN) necessitates node localization, especially if the information to be shared is the location itself, such as in warehousing and information logistics. Trilateration and multilateration positioning methods can be employed in two-dimensional and three-dimensional space respectively. These methods use distance measurements and analytically estimate the target location; they suffer from decreased accuracy and computational complexity especially in the three-dimensional case. Iterative optimization methods, such as gradient descent (GD), offer an attractive alternative and enable moving target tracking as well. This chapter focuses on positioning in three dimensions using time-of-arrival (TOA) distance measurements between the target and a number of anchor nodes. For centralized localization, a GD-based algorithm is presented for localization of moving sensors in a WSN. Our proposed algorithm is based on systematically replacing anchor nodes to avoid local minima positions which result from the moving target deviating from the convex hull of the anchors. We also propose a GD-based distributed algorithm to localize a fixed target by allowing gossip between anchor nodes. Promising results are obtained in the presence of noise and link failures compared to centralized localization. Convergence factor issues are discussed, and future work is outlined.

Keywords: gradient descent, node localization, tracking, distributed averaging, push-sum algorithm, link failure, step size

1. Introduction

Wireless sensor networks are used in a wide range of monitoring and control applications such as traffic monitoring, environmental monitoring of air, water, soil quality or temperature, smart factory instrumentation, and intelligent transportation. The nodes are usually small radio-equipped low-power sensors scattered over an area or volume of a few tens of square or cubic meters, respectively. There is information sharing between sensors and for this information to be meaningful, the nodes or sensors have to be located. Although global positioning systems (GPS) achieve powerful localization, it is costly and impractical to equip each sensor in a WSN with a GPS device. Besides, in many environments such as indoors and forested zones, the GPS signal may be weak or even unavailable. This explains the vast on-going research devoted to efficient localization for WSNs.

Node information may be processed either centrally or in a distributed manner. In centralized localization, distance measurements are collected by a central processor prior to calculation. In distributed algorithms, the sensors share their information only with neighbors but possibly iteratively. Both methods face the high cost of communication, but, in general, centralized localization produces more accurate location information, whereas distributed localization offers more scalability and robustness to link failures.

Node localization relies on the measurements of distances between the nodes to be localized and a number of reference or anchor nodes. The distance measurements can be via radio frequency (RF), acoustic, or ultra-wideband (UWB) signals. Measurements that indicate distance can be time of arrival (TOA), angle of arrival (AOA), or received signal strength (RSS). TOA measurements seem to be most useful especially in low-density networks, since they are not as sensitive to inter-device distances as AOA or RSS. The TOA distance measurements usually correspond to line-of-sight (LOS) arrivals that are hampered by additive noise. The consequent measurement errors can be adequately modeled by zero-mean Gaussian noise with variance σ^2. The inclusion of a mean μ in this Gaussian model may be necessary to account for possible non-line-of-sight (NLOS) arrivals.

Accurate location information is important in almost all real-world applications of WSNs. In particular, localization in a three-dimensional (3D) space is necessary as it yields more accurate results. Trilateration and multilateration positioning methods [1] are analytical methods employed in two-dimensional (2D) and three-dimensional (3D) spaces, respectively. These methods use distance measurements to estimate the target location analytically, and suffer from poor performance, decreased accuracy, and computational complexity especially in the 3D case. More specifically, trilateration is the estimation of node location through distance measurements from three reference nodes such that the intersection of three circles is computed, thereby locating the node as shown in **Figure 1**. Multilateration is concerned with localization in a 3D space in which more than three reference nodes are used [2].

Practically, when distance measurements are noisy and fluctuating, localization becomes difficult. The intersection point in **Figure 1** becomes an overlapped region. With this uncertainty, analytical methods become almost useless and we resort to optimization methods. Iterative optimization methods offer an attractive alternative solution to this problem. The Kalman filter, which is an iterative state estimator, can be used for node localization in case of noisy measurements. However, its computational and memory requirements may not be met adequately by the limited resources of a sensor system, subsequently resulting in poor

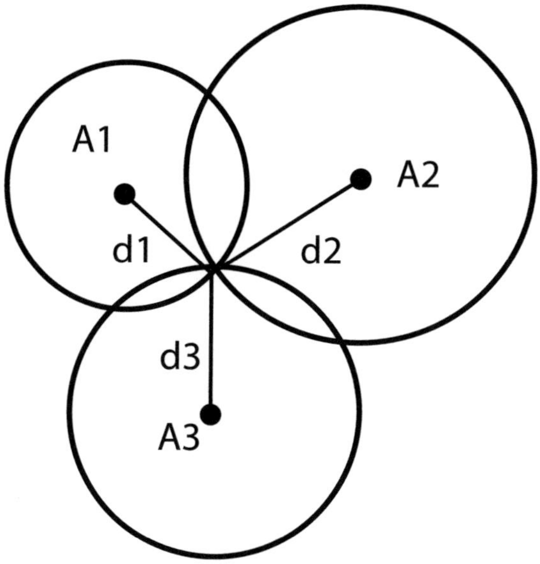

Figure 1. Trilateration.

performance [3]. Thus, the most common iterative optimization method is the computationally efficient gradient descent algorithm, which has been widely dealt with in the literature for the 2D case [4, 5].

This chapter addresses localization in a three-dimensional space of stationary and moving wireless sensor network nodes by gradient descent methods. First, it is assumed that a central processor collects the data from the nodes. TOA measurements will be assumed throughout. An evaluation analysis of the performance of the localization algorithm considered is performed. The effect of measurement noise has also been studied. The work also investigates tracking of moving sensors and proposes a method to counteract some associated problems such as falling into local minima [6]. Second, distributed GD localization will be handled using a proposed gossip-based technique in which anchor nodes exchange data to iteratively compute the positions and gradients locally in each anchor [7]. This distributed method serves to mitigate the effects of noise and link failures.

2. Centralized gradient descent (GD) localization in 3D wireless sensor networks

2.1. Stationary node localization

Localization in 3D space is particularly important in practical applications of WSNs, but many of its aspects remain unexplored as the typical scenario for WSN localization is set up in a 2D plane [8]. In a 3D space, at least four anchor nodes are needed whose locations are known. An estimate of the ith distance d_i, $i = 1, 2, 3, 4$, between the ith anchor node (x_i, y_i, z_i) and the node to be localized (x, y, z) is needed.

The TOA distance measurement technique is assumed. TOA is the time delay between transmission at the node to be localized and reception at an anchor node. This is equal to the distance d_i divided by the speed of light if either RF or UWB signals are used. The backbone of the TOA distance measurement technique is the accuracy of the arrival time estimates. This accuracy is hampered by additive noise and NLOS arrivals. The measurement errors are modeled as additive zero-mean Gaussian noise. The total additive Gaussian measurement noise will be modeled as $N(\mu, \sigma_{\text{NLOS}}{}^2)$, where the letter N denotes the normal or Gaussian distribution, μ is the mean, and σ^2_{NLOS} is the variance, taking into account NLOS as well as LOS arrivals. The occasional inclusion of a mean accounts for the biased location estimate resulting from NLOS errors [9, 10].

To determine the TOA in asynchronous WSNs, two-way TOA measurements are used. In this method, one sensor sends a signal to another that immediately replies. The first sensor will then determine TOA as the delay between its transmission and reception divided by two [10].

Gradient descent iterative optimization in three dimensions results in slower convergence when compared to the 2D case due to tracking along an extra dimension. This is true for all iterative optimization methods. Due to limited exploration of 3D scenarios in the literature, the present work presents practical results relating to the GD localization problem in three-dimensional WSNs. The definition of an objective or error function is normally required for optimization methods whose purpose is to minimize this function to produce the optimal solution. In GD localization, the objective error function is usually defined as the sum of squared distance errors from all anchor nodes. As such, we may write the objective error function as:

$$f(p) = \sum_{i=1}^{N} \{ \left[(x - x_i)^2 + (y - y_i)^2 + (z - z_i)^2 \right]^{1/2} - d_i \}^2 \tag{1}$$

and

$$d_i = c(t_i - t_o) \tag{2}$$

where $p = [x, y, z]^T$ is the vector of unknown position coordinates (x, y, z), t_i is the receive time of the ith anchor node, t_o is the transmit time of the node to be localized, c is the speed of light $(= 3 \times 10^8$ m/s), and N is the number of anchor nodes. The difference $(t_i - t_o)$ is the TOA that can be measured (with measurement noise) in asynchronous WSNs as explained.

Minimization of the objective function produces the optimal solution that is the position estimate of the node to be localized. This problem is solved iteratively using GD as follows:

$$p_{k+1} = p_k - \alpha . g_k \tag{3}$$

where p_k is the vector of the estimated position coordinates, α is the step size, and g_k is the gradient of the objective function given by:

$$g_k = \nabla f(x, y, z) = \left[\frac{\partial f}{\partial x}, \frac{\partial f}{\partial y}, \frac{\partial f}{\partial z}\right]^T \quad (4)$$

If we define the term $B_{k,i}$ as:

$$B_{k,i} = [(x_k - x_i)^2 + (y_k - y_i)^2 + (z_k - z_i)^2]^{\frac{1}{2}} \quad (5)$$

then the three components of the gradient vector at the kth iteration will be:

$$\left.\frac{\partial f}{\partial x}\right|_k = \sum_{i=1}^{N} 2\{B_{k,i} - d_i\}. \frac{(x_k - x_i)}{B_{k,i}} \quad (6)$$

$$\left.\frac{\partial f}{\partial y}\right|_k = \sum_{i=1}^{N} 2\{B_{k,i} - d_i\}. \frac{(y_k - y_i)}{B_{k,i}} \quad (7)$$

$$\left.\frac{\partial f}{\partial z}\right|_k = \sum_{i=1}^{N} 2\{B_{k,i} - d_i\}. \frac{(z_k - z_i)}{B_{k,i}} \quad (8)$$

The initial position coordinates may be chosen to be the mean position of all anchor nodes. The required number of iterations for convergence is a tradeoff between energy consumption, which is critical to WSNs, and the degree of accuracy.

A minimum of four anchor nodes are needed to estimate position in a 3D space. The estimation accuracy increases as a function of the number of anchor nodes. Since the objective function is the sum of the squares of the differences between estimated distances and measured distances, distance measurement errors are squared, too. This problem is countered by weighting distance measurements according to their confidence to limit the effect of measurement errors on localization results [11]. So the objective function accommodating different weights may be expressed as:

$$f(p) = \sum_{i=1}^{N} w_i\{[(x - x_i)^2 + (y - y_i)^2 + (z - z_i)^2]^{1/2} - d_i\}^2 \quad (9)$$

Weighting, however, may result in sub-optimal solutions if only four anchor nodes are used. Since usually there are only a few anchors in a real WSN [12], the use of five anchor nodes is a good choice to achieve better accuracy without undue deviation from real settings.

In a 3D WSN, the error function of Eq. (1) is a 4D performance surface with a global minimum and several local minima. To avoid local minima, the gradient descent must run several times with different starting points, which is expensive computationally. To better visualize the local minima problem, localization in a 2D space is considered to enable performance surface plotting in a 3D space. Three anchors (30, 45), (80, 65), and (10, 80) are chosen with d_i = 32.0156, 83.2166, and 60.0000 corresponding to a point p = (10, 20). Then, plotting the following objective function

$$f(p) = \sum_{i=1}^{3} \{[(x - x_i)^2 + (y - y_i)^2]^{1/2} - d_i\}^2 \tag{10}$$

results in **Figure 2** with azimuth = 90° and elevation = 0°.

The presence of a global minimum at p and a neighboring local minimum can be discerned from **Figure 2**. Therefore, GD search of the minimum along the performance surface often gets trapped in a local minimum especially when tracking a moving node. In the following section, a solution will be presented to solve the local minima problem in a moving sensor localization setting.

2.1.1. Simulation scenario

GD localization in a 3D WSN is simulated in MATLAB. The anchor node locations are chosen at random in a volume of 200 × 200 × 200 m^3. It is assumed that the target node to be localized (whether stationary or moving) has all anchor nodes within its radio range, and that the target node lies within the convex hull of the anchors. The LOS and NLOS measurement noise is assumed to obey a normal distribution $N(\mu, \sigma^2)$. In subsequent simulations, noisy TOA measurements are simulated by adding a random component to the exact value of the time measurement. The latter is readily computed for simulation purposes from knowledge of the exact node position to be localized, the anchor positions, and the speed of light c.

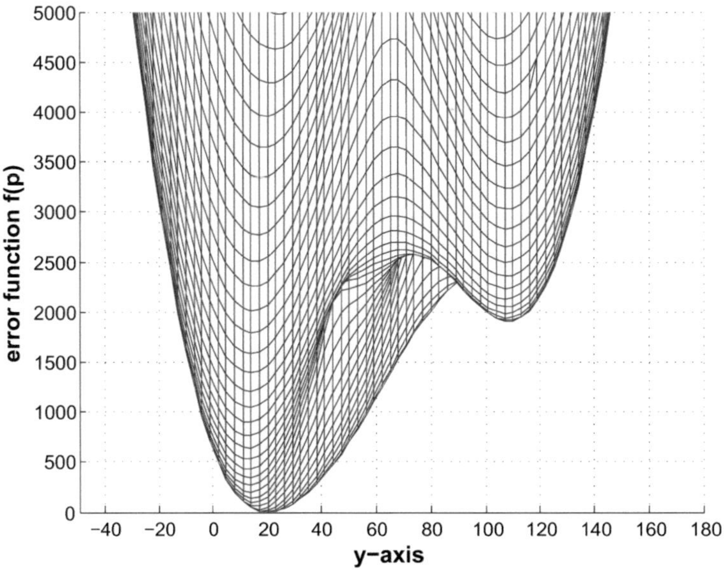

Figure 2. Error function $f(p)$ as a 3D performance surface with 2D anchor nodes (30, 45), (80, 65), and (10, 80) and a global minimum at p = (10, 20). Azimuth = 90° and elevation = 0°.

2.1.2. Simulation results

We first consider four anchor nodes to localize a node of position (60, 90, 60) in the 3D space assuming that the standard deviation (SD) of the zero-mean Gaussian TOA measurement noise, the convergence factor or step size, and the number of iterations to be $SD = 0.001$ μs, $\alpha = 0.25$, and $j = 100$, respectively. The anchor positions are (10, 100, 10), (100, 90, 10), (10, 70, 100), and (100, 80, 100). Simulation results localized the target node as (60.28, 84.02, 58.65). When five anchor nodes are used, they provide an almost ideal target localization of (60.16, 89.64, 60.09). The fifth anchor position is (90, 90, 150). **Figure 3** is a plot of the error function versus the number of iterations for this last case of five anchor nodes. Retaining this scenario, another node (70, 45, 60) is localized as (70.03, 45.16, 59.85). Obviously, any node within the convex hull of the anchor nodes will be almost exactly localized with five anchors.

The results of **Figure 3** are repeated in **Figure 4** taking into account the presence of NLOS arrivals and a greater noise standard deviation. In **Figure 4**, $SD = 0.002$ μs, and $\mu_{NLOS} = 0.006$ μs. A reduction in the localization process accuracy is readily noticed: The point (60, 90, 60) results in a localization of (60.35, 88.97, 59.40). It is also clear from the figure that the solution is biased due to NLOS arrivals.

The issue of energy consumption may appear to disfavor iterative methods compared to analytical methods. This is not the case, however, when the target is moving, since updating would then be must whether iterative or other methods are employed.

2.2. Moving node localization and tracking

GD can be used to track a moving target in real time. The measurement sample interval determines the measurement update rate. A bit of care is required in adjusting the sample

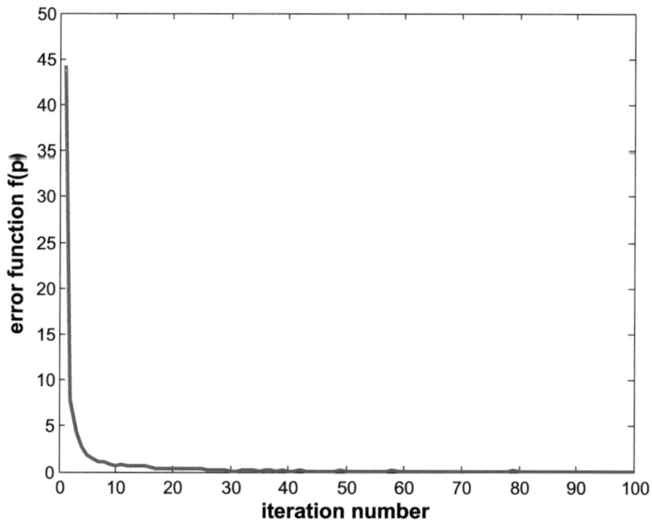

Figure 3. Error function versus the number of iterations when GD localization of a stationary target in 3D space is performed using five anchor nodes. Convergence factor = 0.25, TOA measurement noise SD = 0.001 μs.

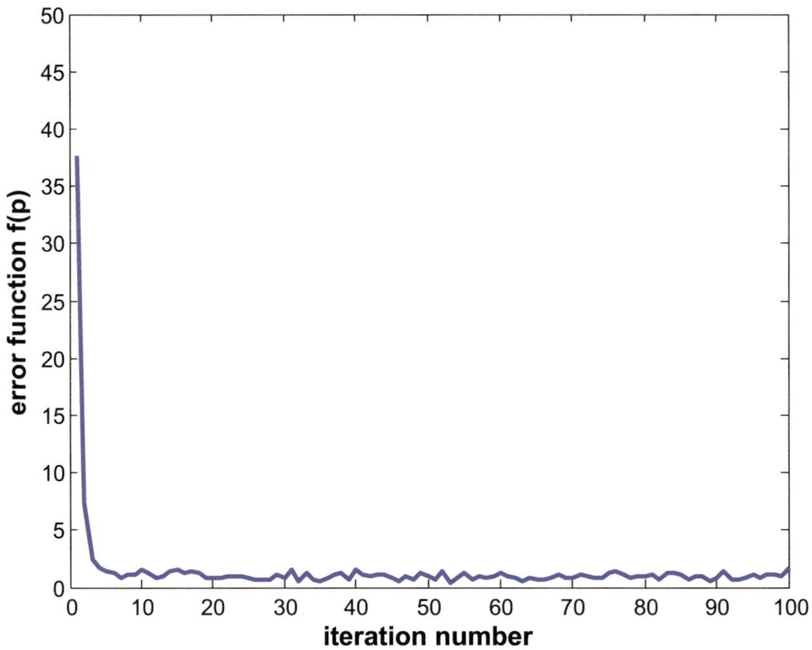

Figure 4. Error function versus the number of iterations when GD localization of a stationary target in 3D space is performed using five anchor nodes. Convergence factor = 0.25, TOA measurement noise SD = 0.002 μs and μ_{NLOS} = 0.006 μs.

interval to avoid conflict with moving sensor velocity and motion models which may be completely unknown [9]. The moving node must provide multiple measurements to the anchors as it moves across space. It has the opportunity to reduce environment-dependent errors as it averages over space. Many computational aspects of this problem remain to be explored [10].

In Refs. [13, 14], the problem of avoiding local minima for moving sensor localization is handled by smart use of available anchors and good initialization. Although these works are also based on minimizing cost functions, they are not general GD algorithms. Moreover, these works require good initial estimation of the target location. It is therefore worthwhile to attempt achieving moving sensor localization without the need to estimate the initial moving target location. As a solution to this problem, we introduce the concept of diversity in the iterative GD localization problem.

The algorithm below is proposed in Ref. [6] to localize a moving sensor in a 3D space with the provision of local minima avoidance. The foreseen success of the proposed method is based on the idea that, as the updated position begins to wander away from the global minimum in the direction of a local minimum, it is highly probable that it will return to the right track if some anchor nodes are replaced. Anchor node replacement results in a consequent change in the performance surface shape and hence local minima positions.

Algorithm 1: Proposed GD localization of a moving sensor [6]

1. Estimate a suitable measurement sample interval or update rate.
2. Cluster available anchor nodes into sets of five nodes each. The number of resulting sets P will be:

$$P = \binom{N}{5} = \frac{N!}{5!(N-5)!}$$

where N is the total number of heard anchor nodes.

3. Randomly draw M sets from P obeying a uniform distribution.
4. Perform M independent gradient descent localization procedures on the moving sensor using these M sets.
5. Iterate the gradient descent algorithm up to the L-th update, and calculate the final $f(p)$ for each of the M sets. Discard the sets that produce $f(p)$ greater than a certain threshold γ. Find the point p with the minimum $f(p)$.
6. Stop the algorithm if the moving sensor tracking halts.
7. Complete the M sets by randomly choosing other sets from P, and repeat steps 4–6 starting with the final position of p that corresponds to the minimum $f(p)$.

The different parameters appearing in Algorithm 1 should be properly chosen. These are M, N, L, and the threshold γ. As discussed in the problem description, N should not be unduly large in practical settings. Assuming that five anchors per set are involved in localization, N must not be much greater especially when the WSN area or volume is limited. As for M, it naturally determines the computational overhead; GD localization must run M times in each round of position estimation. To reduce the amount of computation to a minimum, the choice of M must achieve a tradeoff between computational complexity and sufficient diversity of anchor sets in order to cancel unsuitable candidates and retain functional ones. The threshold γ depends on the specific application and how tolerant the latter is to the final value of the error function $f(p)$. In the simulations, the moderate value of 7 m^2 is used as a default setting. This means that the estimated squared distance error associated with each anchor is (7/5) m^2 on average according to Eq. (1).

As for L, it has been assigned the value 150 iterations in the present simulation settings, which is, however, an ad hoc value that worked for the particular settings under consideration. To ensure accurate tracking, a check on the error function of all running estimations can be performed after each certain interval (e.g. 30 iterations) and then the decision is made whether to proceed or replace the diverging sets.

Applying an iterative optimization algorithm for M subsets of heard anchors, when M can be as large as 20, has been implemented in Ref. [12], albeit without diversity, in the context of least median square (LMS) secure node localization in WSNs to combat localization attacks. Algorithm 1 has been inspired from Ref. [12] by adapting it to:

1. Suit the simpler iterative GD localization algorithm with the aim of local minima avoidance rather than secure localization.

2. Repeat itself with diversity to avoid divergence due to local minima as the target moves along its path.

A final remark concerns the communication overhead; the proposed algorithm does not add to the communication complexity. With each iteration, and after the sensing has been achieved, only one broadcast (communication) of the distance measurement is enough from each of the N anchors. It is in the fusion center that the various combinations of P are sorted out and their associated computations performed.

2.2.1. Simulation results

In the following scenarios, a moving node is tracked and localized. We assume five anchor nodes since this offers the best estimation accuracy. We first illustrate GD tracking of a node moving along a helical path (**Figure 5**). The three dimensions representing the moving target location are given by:

$$\begin{aligned} x &= r \cos \theta \\ y &= r \sin \theta \\ z &= k \theta \end{aligned} \tag{11}$$

The angle θ is continuously increasing and r and k are constants. **Figure 5** shows the moving node helical path and its GD track for values of θ varying from zero to 2π, together with the anchor positions shown as small circles. The anchors are assumed to be in the radio range of the helical trajectory. The constant values r and k are 40 and 20, respectively. Noise-free distance measurements are assumed throughout.

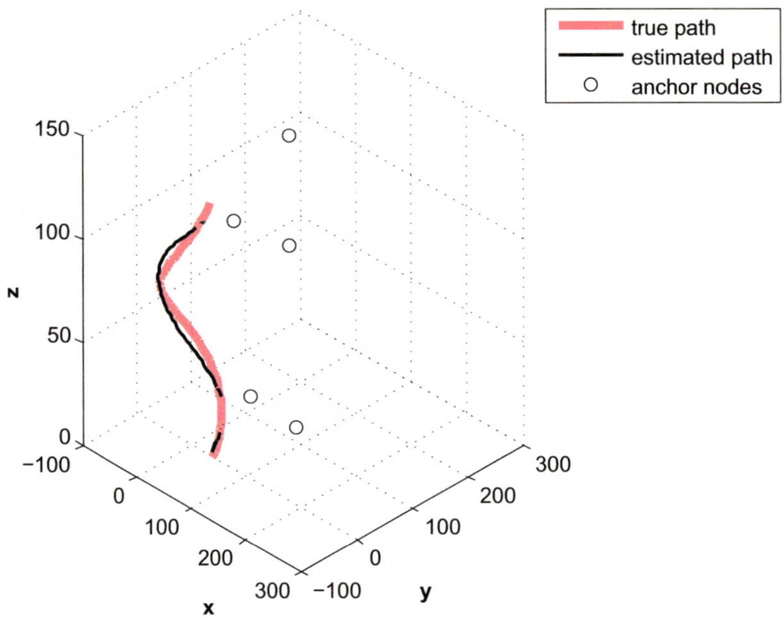

Figure 5. Target node tracking along a helical path.

Next, and to better illustrate the proposed algorithm in Ref. [6] for moving target tracking, and the effect of the various inherent parameter values, a straight line path segment is considered. The details are outlined in the following steps:

a. A target node is moving 0.5 m in each of the three x, y, and z axes in each of 200 steps, which gives a true track distance of 100 m/dimension. The true track is illustrated by the straight line in **Figure 6**. The estimated track begins with an initial point of (50, 50, 50) and converges to the true track for a while but then deviates from it due to the local minima associated with this problem. This deviation is shown clearly in **Figure 6**.

b. The same scenario is repeated except that the track is divided into two segments. The first segment uses the same previous anchor nodes. In the second segment, the anchor nodes have been changed in an attempt to avoid the local minimum and resume tracking the true path. **Figure 7** shows the corrected tracking behavior and the new set of anchor nodes.

c. The proposed method of Algorithm 1 is applied with $N = 7$ resulting in $P = 21$, that is, seven anchor nodes are clustered in 21 sets of five anchor nodes each. M is chosen to be equal to 10 and L equal to 150. The threshold is chosen as $\gamma = 7$. At the 150th update, the final $f(p)$ is calculated for each of the 10 sets. The sets that produce an error function greater than 7 are discarded, and other sets from the remaining 11 sets are chosen to complete the 10 sets starting with the final position of p that corresponds to the minimum $f(p)$. Iterative computations are continued for another 150 updates and the optimum set is

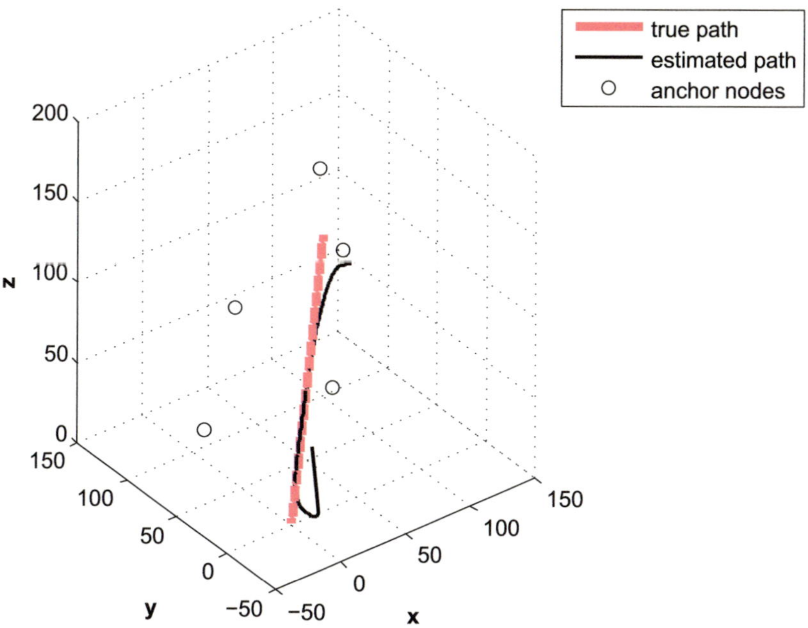

Figure 6. Tracking of a moving sensor in 3D space using iterative GD with initial point (50, 50, 50) and a fixed set of anchor nodes. Convergence factor = 0.1.

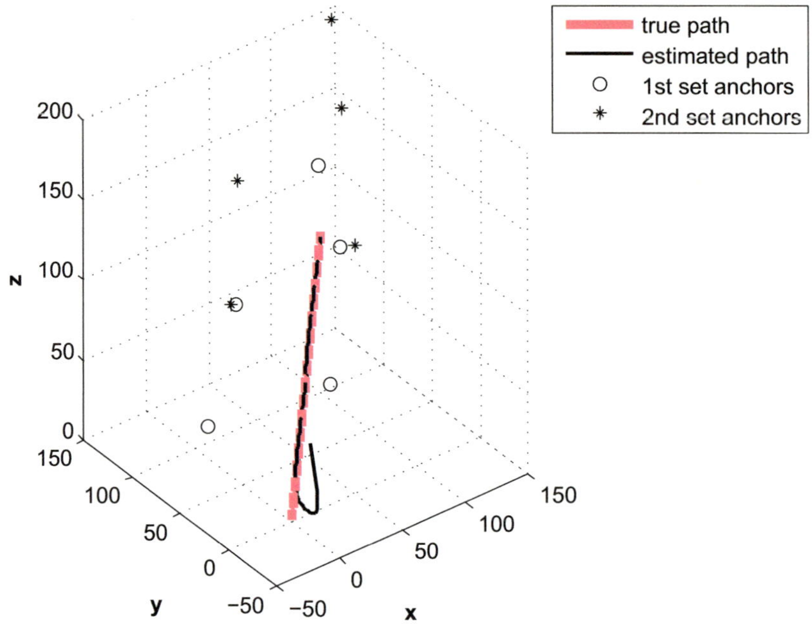

Figure 7. Two-segment true path and track of a moving sensor in 3D space using iterative GD. Initial point is (50, 50, 50). Convergence factor = 0.1.

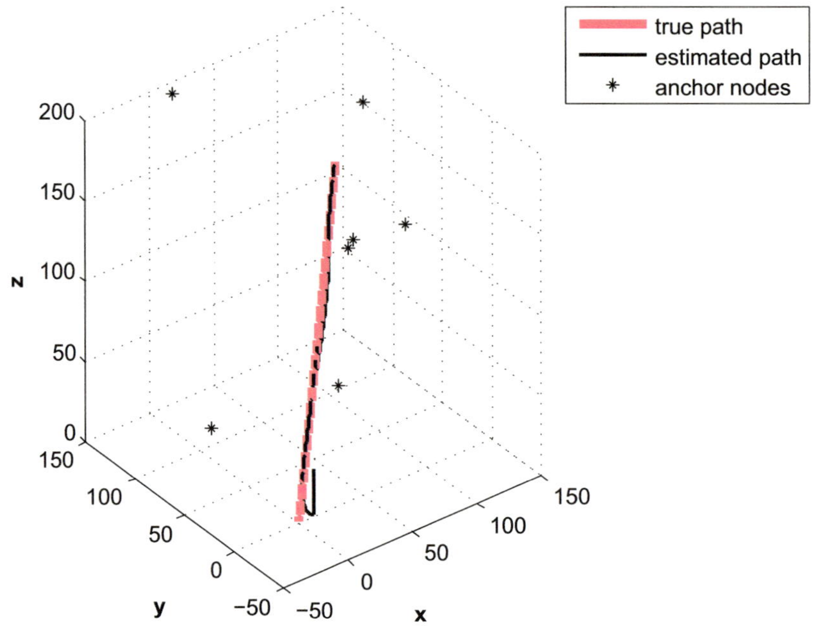

Figure 8. GD tracking of a moving sensor using the proposed algorithm. Initial point is (50, 50, 50). Convergence factor = 0.1.

also found by inspecting the localized point that results in the minimum final $f(p)$. The true and estimated tracks are shown in **Figure 8**. Simulations show that the optimum set of anchor nodes in the first segment (150 iterations) is different from that of the second segment and no local minimum deviation is noticed.

It is worth noting that in the second segment, the first segment unsuccessful sets can be replaced in a deterministic manner rather than randomly, since one would by then have an idea of the location of the moving target. This is especially convenient for WSNs with widely scattered sensors, where sets with nodes that are distant from the moving target and that are likely to contribute to poor localization can be discarded.

Future work may consider introducing distance-measurement noise and studying its effect on the performance of the proposed algorithm. In that case, the final $f(p)$ may not be enough indication of the validity of any certain set of anchors due to noisy measurements. So averaging $f(p)$ of the last 10 iterations of each segment of the estimated path, and for all M running sets, may be considered to obtain a more accurate comparison and a judicious subsequent selection of sets.

3. Distributed gradient descent (GD) localization in 3D wireless sensor networks

In Ref. [7], the authors propose a distributed GD localization method that is robust against node and link failures. The computation of sums is inherent in the GD localization problem and can therefore be made distributed by applying gossip-based distributed summing or averaging algorithms.

It can be seen from Eqs. ((1), (6)–(8)) that, there are four N-term sums that have to be computed in each iteration of the GD localization algorithm. For each of the four sums, each set of variables that constitute each of the N terms is resident in one of the N anchor nodes. This set of variables includes the current tracked or estimated position, the corresponding distance measurement and the location of the anchor node itself. This readily implies the possibility of computing each of the four sums in a distributed manner by sharing information (gossiping) among the anchor nodes. Upon completion of the distributed averaging or summing task, each anchor will possess an estimated value of all four sums, and then Eq. (3) can be computed in each anchor to obtain the estimated position of the node to be localized. This whole process is repeated in each iteration of the GD localization algorithm.

The averaging or summing problem is the building block for solving many complex problems in signal processing. Gossip algorithms [15] are a class of randomized algorithms that solve the averaging problem through a sequence of pairwise averages. In our case, the communicating or gossiping nodes are the anchors, and we assume they are within transmission range of each other. Therefore, a simple gossip-based synchronous averaging protocol called the push-sum (PS) distributed algorithm [15, 16] is used for this application.

3.1. The push-sum gossip-based distributed averaging algorithm

The PS algorithm is iterative and not exact. Therefore, every anchor node will obtain an estimate of the sums that differ slightly from that of the other anchors. The gossiping anchor nodes are assumed to work synchronously. The term "iteration" will be preserved for the GD time step, whereas the term "round" or "PS round" will used to indicate the PS time step. The total number of rounds will be designated as T. With every round t, a weight $\omega(i)$ is assigned to each node i, and initialized to $\omega(i) = 1/N$, where N is the number of anchors. Likewise, a sum $s(i)$ is initialized to $s(i) = x(i)$, where $x(i)$ is the resident summation element in node i. For round $t = 0$, each node i sends the pair $[s(i), \omega(i)]$ to itself, and in each of the remaining rounds $t = 1,\ldots,T$, node i follows the protocol of Algorithm 2:

Algorithm 2: The push-sum algorithm {Pushsum(x_i)} [15, 16]

Input: N and T

1. Initialization: $t = 0$, $s(i) = x(i)$ *and* $\omega(i) = 1/N$ *for* $i = 1, \ldots, N$.
2. **Repeat**.
3. Designate $\{\hat{s}(r), \hat{\omega}(r)\}$ as the set of all pairs sent to node i at round t-1.
4. Compute $s(i) \equiv \sum_r \hat{s}(r)$ and $\omega(i) \equiv \sum_r \hat{\omega}(r)$.
5. At each node i, a target node $f(i)$ is chosen uniformly at random.
6. The pair $[0.5\, s(i), 0.5\, \omega(i)]$ is sent to target nodes $f(i)$ and node i (the sending node itself).
7. $[s(i)/\omega(i)]$ is the estimate of the sum at round t and node i.
8. $t = t + 1$.
9. **until** $t = T$.

Output:

$[s(i)/\omega(i)]$ is the sum at round t and node i.

$\sum_{i=1}^{N} \omega(i) = 1$ and $\sum_{i=1}^{N} s(i) = \textit{the sum}$, at all rounds t.

The number of steps T needed such that the relative error in Algorithm 2 is less than ε with probability at least $(1 - \delta)$ is of order:

$$T(\delta, N, \varepsilon) = O\left(\log_2 N + \log_2 \frac{1}{\varepsilon} + \log_2 \frac{1}{\delta}\right) \tag{12}$$

where T is also referred to as the diffusion speed of the uniform gossip algorithm [15].

3.2. Distributed GD localization in WSNs

The PS distributed averaging method of Algorithm 2 is considered as scalar version. It can be extended to a vector version [17] where nodes (anchors) exchange vector messages that are summed up element-wise. This concept readily conforms to our proposed distributed GD localization method in which we have to compute four sums in each iteration as in Eqs. ((1), (6)–(8)).

At the kth iteration and in the ith anchor node, there reside $f(p_k)|_i$, $\frac{\partial f}{\partial x}|_{k,i}$, $\frac{\partial f}{\partial y}|_{k,i}$, and $\frac{\partial f}{\partial z}|_{k,i}$ which can be considered the four elements of the vector.

The core idea of our distributed GD localization algorithm is that, for each outer gradient iteration, a series of inner rounds reach consensus on each of the four N-term sums.

3.2.1. Simulation results

The GD localization problem in a 3D space is simulated in MATLAB as in Ref. [7]. Four anchor node locations are chosen in a volume of $100 \times 100 \times 100$ m^3. It is assumed that the target node to be localized has all anchors within its radio range. The same four anchors given in the simulation results of Section 2 are used. The targeted node is (60, 90, 60). Error-free TOA measurements are assumed, and centralized localization is first performed with $N = 4$, $\alpha = 0.25$ and p_o = (50, 50, 50). After 100 iterations, it is found that the error function is 0.748 and the localized point is (60.1, 84.1, 58.8) which is very close to the targeted node.

Treating the order in Eq. (12) as an exact value, we set the number of rounds of the PS algorithm (Algorithm 2), T, for a number of nodes N, equal to

$$T = \log_2\left(\frac{N}{\delta\varepsilon}\right) \tag{13}$$

Note that $\delta\varepsilon = 2^{-12}$ is obtained when we set $\delta = \varepsilon = 2^{-6} \approx 0.0157$. Substituting these values in Eq. (13), we find that $T = 14$ PS rounds when $N = 4$. Clearly, this implies that we may expect a relative error $\varepsilon \leq 0.0157$ with probability higher than 0.9843 in the PS algorithm. The final accuracy in the estimated localization corresponds to the accuracy level ε set in the PS algorithm [16]. Thus, from such estimated values of ε and $(1 - \delta)$, it can be deduced that the accuracy of our distributed localization algorithm is almost equivalent to that of centralized GD localization in the absence of noise and link failures.

Distributed algorithms are robust against network failures, or, typically, link failures. The latter arise due to many reasons such as channel congestions, message collisions, moving nodes, or dynamic topology [18]. Link failures can be modeled by the absence of a bidirectional connection between two nodes. All nodes operate in synchronism. At each time step, some percentage of the links between anchor nodes is randomly removed. The missing links may differ every time step since they are programmed to be randomly chosen, but their number remains fixed for each run of the code, and ensemble averaging over 100 trials is performed in each run. **Figure 9** demonstrates the robustness of the proposed distributed algorithm. Even if we lose up to 50% of the links in every time step, the algorithm is still comparatively accurate. This is illustrated by **Figure 9a** and **b** which are plots of the error function versus iteration number in the presence of link failures. For $N = 4$, the number of available links is 6, and losing three (50%) of which results in a final localized target point of (59.8, 82.8, 58.4) with an error function of 0.9 when $\alpha = 0.25$ [7].

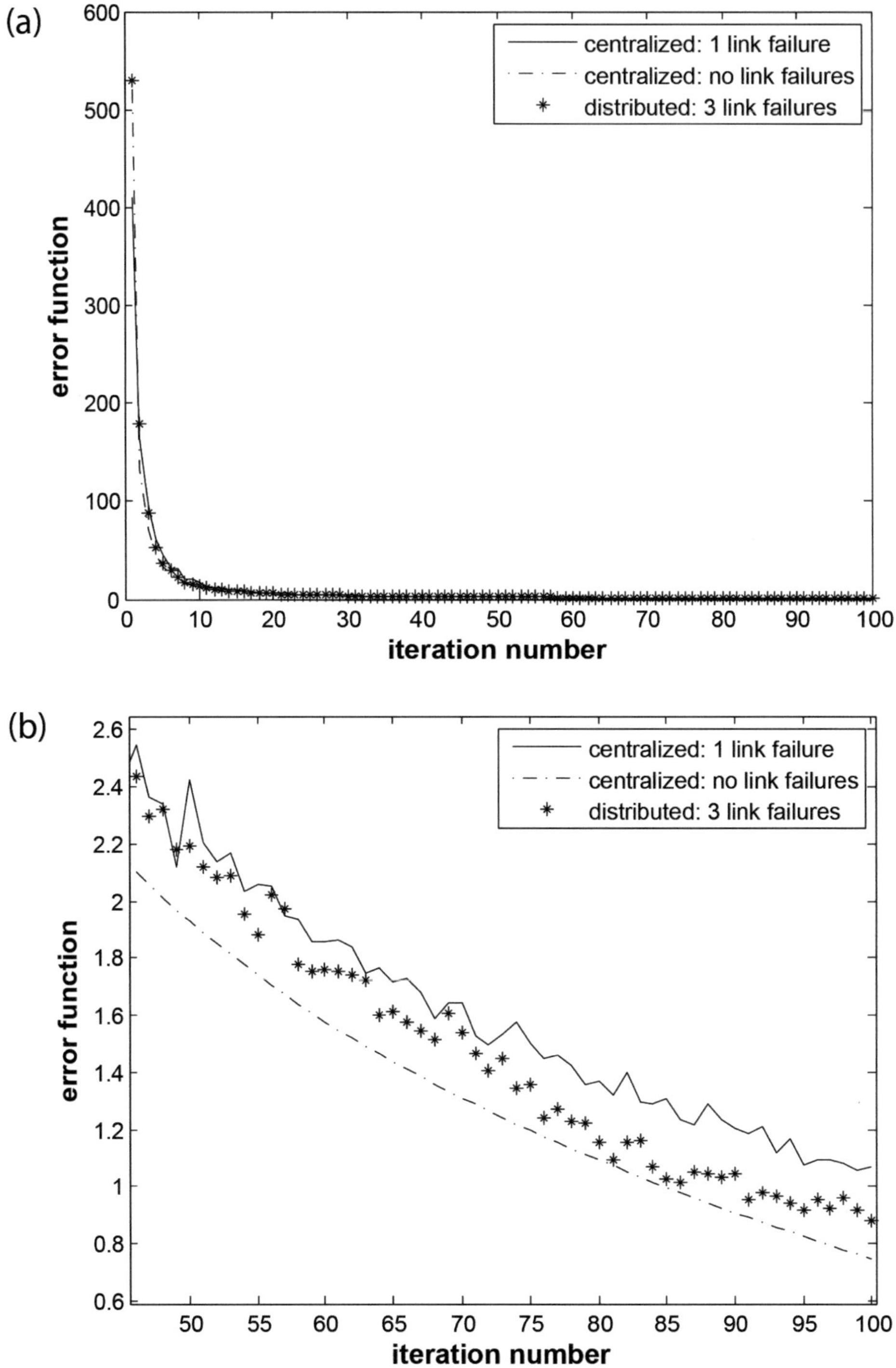

Figure 9. (a) Error function versus iteration number for centralized and proposed distributed GD localization for different cases of link failure conditions, $\alpha = 0.25$. (b) A close view of **Figure 9a** demonstrating the comparative performance of the different centralized and proposed distributed localization algorithms.

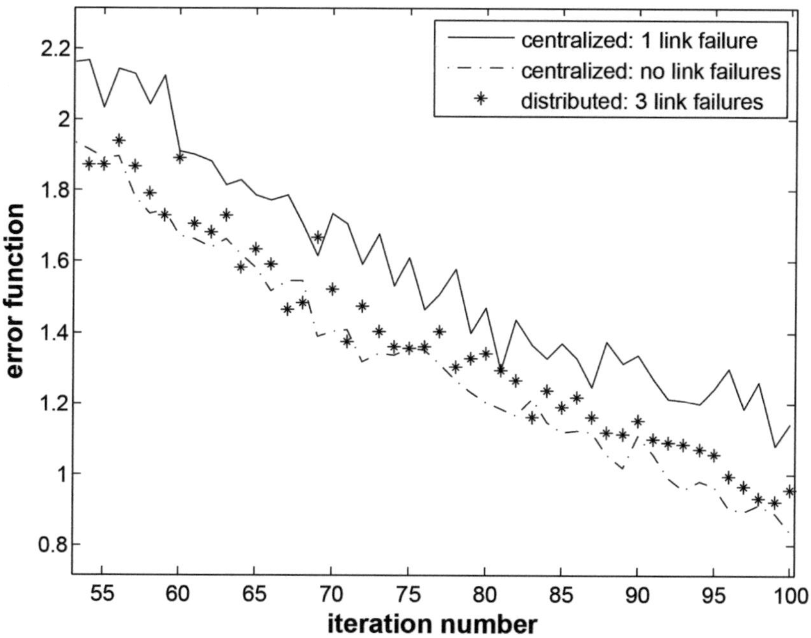

Figure 10. Comparative performance of the different centralized and proposed distributed localization schemes. $\alpha = 0.25$. TOA measurement noise SD = 0.5 ns.

For the purpose of comparison with centralized GD localization, we find that one link failure (25% of available links) isolates the corresponding node from the fusion center, and we have only three anchors to compute the target position, though randomly chosen in every time step. After ensemble averaging, the localized point is (60.1, 82.4, 58.6) and the error function is 1.0, again when $\alpha = 0.25$. This accuracy and in fact, even slightly better is achieved with the distributed scenario of four anchors and three link failures (50% of available links), which clearly shows the advantage of our proposed distributed localization algorithm over its centralized counterpart. The only disadvantage is that in every iteration, we must allow for a delay of 14 PS rounds (T).

The simulations are repeated for noisy TOA measurements as shown in **Figure 10**. Gaussian measurement noise with zero-mean accounting for LOS arrivals only is assumed, and the SD is chosen to be 0.5 ns. This results in a distance error of 15 cm when UWB signals are used for sensing. The resulting plots are noticeably noisier than those of **Figure 9**, but are obviously interpreted in the same way as the noise-free cases. That is, the proposed distributed algorithm with three link failures (50% of links) performs better than the centralized algorithm with one link failure (25% of links) [7].

4. Step size considerations

The fixed step size in this work should be chosen carefully; a too large step size would affect the performance advantage of the proposed distributed localization algorithm as well as the centralized one, whereas a small step size would increase the error function. It is worth mentioning that there are instances in the literature on distributed GD localization algorithms where only the optimal step size is computed in a distributed manner [19, 20] rather than the GD sums in the present work. In Refs. [19, 20], the optimization of the step size in each iteration depends on the node positions and gradients. The optimization method is called the Barzilai-Borwein or simply BB method [21], in which the step size is updated at each iteration using the estimated target position and gradient vectors of the current and past time iterations.

The BB method cannot be applied successfully to our distributed GD localization under consideration [7], that is, by updating α at each iteration and in each anchor. Applying the BB method yields favorable results that are superior to those with fixed step size only in the cases of centralized localization, and distributed localization in the absence of link failures which is an ideal situation not found in practice. The reason is obvious since, in our work, the gradient components are found through gossiping among anchors and become, therefore, greatly affected in case of link failures causing the BB method to result in pronounced sub-optimality in the computation of α at each iteration and in each anchor. This conclusion was arrived at in Ref. [22], where the above situation was simulated and the BB method tested when applied to GD localization in WSNs. Linearly-varying step sizes are shown in Ref. [22] to have the best performance, as they do not involve gradient computations.

5. Recapitulation and future trends

The problem of sensor localization in a 3D space by the method of gradient descent has been investigated and solutions are presented to some impediments that are associated with the moving sensor case, namely, the local minima problem [6]. The proposed method considers all possible combinations of a certain chosen number of anchor nodes from a larger set of available anchors. The foreseen success of the proposed method stems from the fact that a deviating estimated path toward a local minimum is almost certain to return to the right track if some anchor nodes are replaced. This is true since anchor node replacement entails a change of the shape of the performance surface along with different local minima positions. The anchor nodes placement is made uniformly random as the true track of the moving sensor to be localized is unpredictable, and it is performed periodically. The simulation results demonstrate the success of this method. The advantage gained is at the expense of increased computational requirements, and the proposed method also necessitates faster data processing in order to perform accurate moving sensor localization in real time.

In Ref. [7], the GD localization algorithm in WSNs in a 3D space was combined with PS gossip-based algorithms to implement a distributed GD localization algorithm. The main idea is to compute the necessary sums by inter-anchor gossip. The method compared favorably with the centralized version as regards convergence, accuracy, and resilience against noise and link failures. Our simulation results demonstrate that centralized processing with four anchors and one link failure (25% of the links) introduce a localization error comparable to (and even slightly greater than) that introduced by the proposed distributed processing method with three link failures (50% of the links). This is achieved when the number of PS rounds is suitably selected.

Despite the inevitable degradation of performance in case of noisy TOA measurements, the proposed distributed method retains its advantages over centralized processing with proper selection of the GD step size and number of PS rounds. It is therefore evident that resort to distributed techniques such as the proposed distributed GD localization algorithm [7] ensures robustness against link failures even in the presence of noisy TOA measurements, eliminates the need for a computationally-demanding central processor, and avoids a possible communication bottleneck at or near the fusion center [10].

As a future trend, compressive sensing (CS) or random sampling can be implemented to track a moving node in a centralized WSN using the iterative GD algorithm resulting in remarkable energy efficiency with tolerable error [23]. Moreover, an efficient approach for (pseudo-) random sampling via chaotic sequences that has first appeared in Ref. [24] could initiate further investigation of CS concepts via chaos theory and the possibility of their application to WSN moving node tracking.

Author details

Nuha A.S. Alwan[1]* and Zahir M. Hussain[2,3]

*Address all correspondence to: n.alwan@ieee.org

1 College of Engineering, University of Baghdad, Baghdad, Iraq

2 College of Computer Science and Mathematics, University of Kufa, Najaf, Iraq

3 School of Engineering, Edith Cowan University, Joondalup, Australia

References

[1] Zhang L, Tao C, Yang G. Wireless positioning: Fundamentals, systems and state of the art signal processing techniques. In: Melikov A, editor. Cellular Networks—Positioning, Performance Analysis, Reliability. InTech; Rijeka, Croatia. 2011. pp. 3–50. ISBN: 978-953-307-246-3

[2] Alrajeh NA, Bashir M, Shams B. Localization techniques in wireless sensor networks. International Journal of Distributed Sensor Networks. 2013;**2013**:9. DOI: 10.1155/2013/304628

[3] Shareef A, Zhu Y. Localization using extended Kalman filters in wireless sensor networks. In: Moreno M, Pigazo A, editors. Kalman Filter: Recent Advances and Applications. I-Tech; Rijeka, Croatia. 2009. pp. 297–320. ISBN: 978-953-307-000-1

[4] Qiao D, Pang GKH. Localization in wireless sensor networks with gradient descent. In: Proceedings of the IEEE Pacific Rim Conference on Communications, Computers and Signal Processing (Pac-Rim); 23 August 2011; Victoria, BC, Canada. IEEE; 2011. pp. 91–96

[5] Garg R, Varna AL, Wu M. Gradient descent approach for secure localization in resource constrained wireless sensor networks. In: International Conference Acoustics, Speech and Signal Processing (ICASSP); 14 March 2010; Dallas, TX, USA. IEEE; 2010. pp. 1854–1857

[6] Alwan N AS, Mahmood AS. On gradient descent localization in 3-D wireless sensor networks. Journal of Engineering. 2015;**21**:85–97

[7] Alwan NAS, Mahmood AS. Distributed gradient descent localization in wireless sensor netowrks. Arabian Journal for Science and Engineering. 2015;**40**:893–899. DOI: 10.1007/s13369-014-1552-2

[8] Wang J, Ghosh RK, Das SK. A survey on sensor localization. Journal of Control Theory Applications. 2010;**8**:2–11. DOI: 10.1007/s11768-010-9187-7

[9] Gustafsson F, Gunnarsson F. Mobile positioning using wireless networks. IEEE Signal Processing Magazine. 2005;**22**:41–53. DOI: 10.1109/MSP.2005.1458284

[10] Patwari N, Ash JN, Kyperoutas S, Hero AOIII, Moses RL, Correal NS. Locating the nodes. IEEE Signal Processing Magazine. 2005;**22**:54–69. DOI: 10.1109/MSP.2005.1458287

[11] Kwon YM, Mechitov K, Sundresh S, Kim W, Aga G. Resilient localization for sensor networks in outdoor environments. In: The 25th IEEE International Conference on Distributed Computing Systems (ICDCS 2005); 6-10 June 2005; Columbus, OH, USA. IEEE; 2005. pp. 643–652

[12] Li X, Hua B, Shang Y, Xiong Y. A robust localization algorithm in wireless sensor networks. Frontiers of Computer Science China. 2008;**2**:438–450. DOI: 10.1007/s11704-008-0018-7

[13] Agarwal A, Daume HIII, Phillips JM, Venkatasubramanian S. Sensor network localization for moving sensors. In: Proceedings of the 12th International Conference on Data Mining Workshops (ICDMW 2012); 10 December 2012; Brussels: IEEE; 2012. pp. 202–209

[14] Agarwal A, Phillips JM, Venkatasubramanian S. Universal multidimensional scaling. In: Proceedings of the 16th International Conference on knowledge discovery and data mining (ACM SIGKDD 2010); 25-28 July 2010; Washington, DC. ACM; 2010. pp. 1149–1158

[15] Kempe D, Dobra A, Gehrki J. Gossip-based computation of aggregate information. In: Proceedings of the 44th Annual IEEE Symposium on Foundations of Computer Science; 11-14 October 2003; Cambridge, MA. IEEE; 2003. pp. 482–491

[16] Dumard C, Riegler E. Distributed sphere decoding. In: International Conference on Telecommunications (ICT'09); 25-27 May 2009; Marrakech. IEEE; 2009. pp. 172–177

[17] Strakova H, Gansterer WN. A distributed Eigensolver for loosely coupled networks. In: The 21st Euromicro International Conference on Parallel, Distributed and Network-Based Processing (PDP); 27 Feb-1 March 2013; Belfast. IEEE; 2013. pp. 51–57

[18] Sluciak O, Strakova H, Rupp M, Gansterer WN. Distributed Gram-Schmidt orthogonalization based on dynamic consensus. In: The 46th Asilomar Conference on Signals, Systems and Computers; 4-7 November 2012; Pacific Grove CA. IEEE; 2012. pp. 1207–1211

[19] Calafiori GC, Carlone L, Wei M. A distributed technique for localization of agent formations from relative range measurements. IEEE Transactions on Systems, Man, and Cybernetics—Part A: Systems and Humans. 2012;**42**:1065–1076. DOI: 10.1109/TSMCA.2012.2185045

[20] Calafiori G, Carlone L, Wei M. A distributed gradient method for localization of formations using relative range measurements. In: Proceedings of the 2010 IEEE International Symposium on Computer-Aided Control System Design (CACSD'10); 8-10 September 2010; Yokohama. IEEE; 2010. pp. 1146–1151

[21] Barzilai J, Borwein JM. Two point step size gradient methods. IMA Journal of Numerical Analysis. 1988;**8**:141–148. DOI: 10.1093/imanum/8.1.141

[22] Alwan NAS. Adaptive step sizes for gradient descent localization in wireless sensor networks. International Journal of Information and Communication Technology Research. 2016;**6**:1–7

[23] Alwan NAS, Hussain ZM. Compressive sensing for localization in wireless sensor networks: An approach for energy and error control. Submitted for publication. 2017

[24] Nguyen LT, Phong DV, Hussain ZM, Huynh HT, Morgan VL, Gore JC. Compressed sensing using chaos filters. In: Australasian Telecommunication Networks and Applications Conference (ATNAC 2008); 7-10 December 2008; Adelaide, SA. IEEE; 2008

Chapter 4

A Novel Hybrid Methodology Applied Optimization Energy Consumption in Homogeneous Wireless Sensor Networks

Plácido Rogerio Pinheiro,
Álvaro Meneses Sobreira Neto,
Alexei Barbosa Aguiar and
Pedro Gabriel Calíope Dantas Pinheiro

Additional information is available at the end of the chapter

http://dx.doi.org/10.5772/intechopen.70207

Abstract

A wireless sensor network's lifetime is influenced directly by the sensors power management that composes the network. The models applied to the problem aims to optimize the energy usage managing the sensors activation in time intervals, activating only the minimum number of sensors respecting the coverage and connectivity restrictions. However, this problem's class has a significant computational complexity and many applications. It is necessary to implement methodologies to find the optimal solution, increasing the network's size, becoming closer to the real ones. This research's objective is to present a method based on a Partition Heuristic aggregating the Generate and Solve method, improving the results, and increasing the network's instances size, while maintaining the flexibility and reliability when applied to the homogeneous wireless sensors networks with coverage and connectivity restrictions.

Keywords: homogeneous models, wireless sensor network, optimize power consumption, Integer linear programming, hybrid methodology, random key genetic algorithm

1. Introduction

Wireless sensors networks (WSN) was initially used to monitor many phenomena, like temperature, atmospheric pressure, humidity, light, sound volume, solar radiation, and many others. These networks are standard, used in environments with challenging access and hostile. Due to this, methods that optimize the network power consumption to extend the network lifetime are always desired. The network's sensor nodes are typically deployed in the sensed area randomly due to the difficult local access. In these scenarios, some nodes might detect the same field or approximately the same area of another node. This characteristic can lead to redundant sensed data and the premature energy depletion. The solution proposed by [1, 8, 9] is created different network configurations associated with a time interval. In addition to the creation, of many time intervals, it is possible to activate only the necessary sensors to maintain the sensed coverage and connectivity each time interval. Furthermore, the integer linear programming model was proposed by [1, 12] to optimize the energy consumption considering the coverage and connectivity restriction, by allocating the necessary sensors to a pre-defined number of time intervals. As this problem is NP-Complete, it is complexity increases exponentially as the network number of sensors, time intervals, and demand increase.

The solution gets impracticable for the real system sizes. To get closer from real systems, an integrative collaboration of genetic algorithms and integer linear programming tries to merge their high points and has offered significant results improvements [3, 7]. However, its original implementation is shown some deficiencies which limit its performance, being one of them the density explosion. The correct use of a novel hybrid methodology to deal with a class of problem can show good results. The method mainly consists in reducing a problem, typically a high complexity issue, that the time to solve it is infeasible big, into subproblems.

In the wireless sensor network, the genetic algorithm (GA) is used as the Reduced Instances Generator. It is responsible for generating the sub-problems and evolving its population to find the good feasible solution with its value as close as possible from the best solution to the original problem. Although the GA is an excellent approach to solving complex problems in less time, the use of the pure GA might have problems to maintain the feasibility of solutions and execution time. In fact, some operators might be applied to control the chromosome's formation during the crossover operation. Adaptations can be made in the default GA to surpass this problem. The random-key genetic algorithm (RKGA) and biased random-key genetic algorithm (BRKGA) are adaptations of the default GA. They use random generated keys arrays to assist the operation, changing the default GA mutation method. In this work, the BRKGA was adapted attending the WSN problems characteristics. Hybrid approaches are often used to reduce the time spend to solve a problem when a network becomes closer to a real scenario; it finds a difficult time to deal with density explosion. So, the subproblems generated by the genetic algorithm alone (GA and BRKGA) aren't enough to maintain a good solution. Thus, a Partition heuristic is implemented in this problem to enable the more complexes scenarios executions. This new heuristic consists of dividing the whole network, generating sub-networks. This heuristic separates the network area to reduce the complexity. This division is made cautiously not to affect the results on the main problem. After the division, the hybrid methodology is applied over each sub-network. To reconstruct the original problem, a light algorithm, typically a tight one, must be implemented to acquire the first network's solution.

A dynamic model in integer linear programming with restrictions connectivity and coverage applied to the wireless sensor networks had been implemented in this work [4–7]. Moreover, Section 2 is given an outline about wireless sensor network. More details of the models in homogeneous wireless networks are provided is Section 3. On the other hand, in Section 4 provides more information on the applied methodologies. Moreover, in Section 5 the computational results obtained using the heuristic described the integer linear programming model implemented in this work. Finally, Section 6 presents conclusions and future works.

2. The wireless sensor network

The sensor nodes are small devices capable of measure different phenomena. They also can wirelessly communicate to each other to send the sensed data to another instrument. This other appliance, called sink, is a more powerful device responsible for receiving and processing the data from the sensor nodes to finally send it to a server computer. In the same network, more than one sink can be deployed. Furthermore, it has a significant impact on the network topology. A typical sensor node is formed mainly of four components, namely the processing module, the battery, the transceiver, and the sensor module as described [11]. The processor module is responsible for processing all the data sensed but in a simplified way in the microprocessor. The sensor nodes use a small operational system to manage its essential features. Provide power to the sensor; the battery must be present in it. Its lifetime depends on many aspects, like charge capacity and electric current necessary to make the node's components work. The transceiver is a device that sends and receives all the data through radio-frequency propagation. They frequently use public frequency bands to transmit the data. Consequently, external devices might interfere with the network communication. In the same way, communications between sensors nodes of the same network, when there are many close from each other might interfere among themselves, so, the smaller the number of active sensors in the system, more reliable the communication in the network will be.

The sensors coverage radius is an issue that must be finely calibrated by the network manager before it is deployment [14]. For each sensor that should be monitored, a different coverage radius is necessary. On the other hand, this happens because of the phenomenon's variation ratio. For some phenomena, a large coverage space is applicable without denigrating the obtained data quality. Otherwise, some other events need a smaller coverage radius due to theirs high variation and significance on small areas as defined in [15]. The main impact on energy consumption is that the more sensors needed to cover the whole sensed area, the bigger is the current consumed across the network. So, a well-calibrated coverage radius is a critical aspect to be considered. Considering a minimum number of sensor nodes was used to cover a given and often unlimited area of observation a very unreliable and unstable network used to cover a given and often unlimited area of consideration. Furthermore, this is because the coverage areas of the redundant sensor nodes overlap too much, giving birth to redundant data.

3. Model in homogeneous wireless networks

The solution proposed by [7] is to create different schedules, each one associated with a time interval that activates only the set of sensor nodes necessary to satisfy the coverage and connectivity restrictions. The employment of different schedules keeps from occurring the premature starvation from some of the nodes, bringing about a more homogeneous models energy consumption level across the whole network. Moreover, this is provided because the alternation of active nodes among the schedules is often a design outcome, as it optimizes the overall network energy consumption taking into account all time intervals, coverage, and connectivity restrictions. To accurately model the homogeneous wireless sensors networks setting some previous remarks:

- A demand point is a geographical moment in the region of monitoring where one the phenomenon is sensed. Considering the distribution of points across the area of control can be regular, like a grid and random in nature. At least one sensor must be active at a given moment to sense each demand point. Such restriction is implemented in the model.
- Commonly encountered, the sensors are connected with coverage areas that cannot be estimated accurately. To reduce fundamental parts modeling, we assume open areas without obstacles. Moreover, we think a circular coverage area. On the other hand, the coverage radius is determined by the spatial variation of the sensed phenomenon. The radio-frequency propagation in real wireless sensors networks is also irregular in nature.

The energy consumption is the electric current drawn by a circuit in each period. In what follows, the constants, variables, objective function and restrictions of the integer linear programming model Applied Energy Consumption in homogeneous is present in a step-by-step manner.

3.1. Constants

It presents sets aiming to outline the homogeneous model that stand for the constants.

S	Set of sensors
D	Set of demand points
M	Set of sinks
T	Set of n scheduling periods
AD_{ij}	Matrix of arcs ij, $i \in S$, $j \in D$ *which indicate that* sensor i can cover to demand points j
A_{ij}	Matrix of arcs ij, $i \in S$, $j \in S \cup M$ that interconnects sensors
EB_i	Accumulated battery charge for sensor $i \in S$
EA_i	Power needed to activate the sensor node $i \in S$
EM_i	Power required to maintain a sensor $i \in S$ active for a period
ET_{ij}	Power required for transmitting data from the sensor $i \in S$ to sensor $j \in S$.
ER_i	Power needed in the reception of data for sensor $i \in S$
EH	Penalty applied when any sensor does not cover a demand point in any time interval.

3.2. Variables

The following decision variables are necessary to model the coverage and connectivity problem in wireless sensors networks.

x_{tij} If sensor $i \in S$ covers demand point $j \in D$ in period $t \in T$, *is assigned the value* 1, *otherwise the value* 0

z_{tlij} If arc *ij* belongs to the route from sensor $l \in S$ to a sink in period $t \in T$, *is assigned the value* 1, *otherwise the value* 0

w_{ti} If sensor $i \in S$ was activated in period $t \in T$ for at least on phenomenon, *is assigned the value* 1, *otherwise the value* 0

y_{ti} If sensor $i \in S$ is activated in period $t \in T$ *is assigned the value* 1, *otherwise the value* 0

h_{tj} If any sensor does not cover demand point $j \in D$ in period $t \in T$ *is assigned the value* 1, *otherwise the value* 0

e_i Electrical charge consumed by sensor $i \in S$ considering all time periods.

3.3. Objective function

The objective function (1) optimize the total energy used by the sensors all time periods. The second term penalizes the existence some uncovered demand points.

$$\min \sum_{i \in S} e_i + \sum_{t \in T} \sum_{j \in D} EH_{tj} h_{tj} \tag{1}$$

3.4. Restrictions

The constraint (2) imposes the activation of at least one sensor node i to cover the request point j in period t. Otherwise, the variable that indicates the penalty for not covering the request point is activated. Additionally, allows the model not to result in an unfeasible solution in case there is any point of demand that cannot be covered by any sensor node.

$$\sum_{i \in S} \sum_{j \in D} AD_{ij} x_{tij} + h_{tj} \geq 1, \forall j \in D, \forall t \in T \tag{2}$$

Moreover, the restriction (3) indicates to the model that maintenance energy is being consumed. The variable y indicates whether a sensor that monitors the demand point j is active in the interval t.

$$x_{tij} \leq y_{ti}, \forall i, j \in S, \forall t \in T \tag{3}$$

According to the flow conservation principle applied to the connectivity issue, if there is an incoming route to a sensor node, there should be an outgoing route from this same sensor node. Furthermore, this restriction (4) imposes, setting an outgoing route from the sensor node j to sensor node k if there is already an incoming route from sensor node i to the sensor node j.

$$\sum\nolimits_{i\in(S-\{j\})} a_{ij}z^{t}_{lij} - \sum\nolimits_{k\in(SUM-\{j\})} a_{jk}z_{tljk} = 0, \forall j, l \in S, \forall t \in T \tag{4}$$

If there is an active sensor, then there must be a path starting from it, as indicated in restriction (5).

$$\sum_{k\in(SUM-\{l\})} A_{ij}z_{tljk} = y_{ti}, \forall i \in S, \forall t \in T \tag{5}$$

On the other hand, the constraint (6) is necessary to create a path that reaches a sink if a sensor is active.

$$\sum_{i\in S}\sum_{j\in M} A_{ij}z_{tlij} = y_{ti}, \forall t \in T, \forall i \in S \tag{6}$$

Furthermore, in restriction (7), if there is an outgoing route passing through sensor node i, then this sensor node should be active.

$$A_{ij}z_{tlij} \leq y_{tj}, \forall j \in S, \forall l, i \in (S - \{k\}), \forall j, \forall t \in T \tag{7}$$

All data sensed must attain a sink node. On the other hand, the quantity sensors node has no direct connectivity to a sink node. Furthermore, others sensor nodes might be activated just to turn viable the route to the sink. Moreover, with restriction (8) if there is an incoming route passing through sensor i, then this sensor has to be active.

$$A_{ij}z_{tlij} \leq y_{ti}, \forall j \in S, \forall l, i \in (S - \{j\}), \forall t \in T \tag{8}$$

The total energy consumed by a sensor node is the sum of the parcels in the restriction (9).

$$\sum\nolimits_{t\in T}\left(EM_i y_{ti} + EA_i w_{ti} + \sum\nolimits_{l\in(S-\{i\})}\sum\nolimits_{k\in S} ER_i z_{tlki} + \sum\nolimits_{l\in S}\sum\nolimits_{j\in(SUM)} ET_{ij}z_{tlij}\right) \leq e_i, \forall i \in S \tag{9}$$

The maintenance energy is attributed when the sensor is active for any reason. The activation energy is summed only when there was an effective activation through time intervals. The reception and transmission energy are specified when there are incoming and outgoing routes respectively passing from a sensor node. Additionally, the sum has been less or equals than the battery's energy. Additionally, the restriction (10) enforces that each sensor node should consume at most the capacity limit of its battery.

$$0 \leq e_i \leq EB_i, \forall i \in S \tag{10}$$

Also, if a sensor is active in the first time interval, it means it consumed energy to activate. The w variable indicates this activation. In addition to, the variable's value is set to 1. By the other hand, if the sensor is kept off in the first-time interval, the value is set to 0. Restriction (11) ensures that.

$$w_{0i} - y_{0i} \geq 0, \forall i \in S \tag{11}$$

http://dx.doi.org/10.5772/intechopen.70207

Also, in restriction (12) the sensor's past and current activation states are compared. If the sensor node was active from period t − 1 to period t, then w is set to 1, 0 otherwise.

$$w_{ti} - y_{ti} + y_{(t-1)i} \geq 0, \forall i \in S, \forall t \in T, \ t > 0 \quad (12)$$

Finally, the restriction (13) only indicates that the decision variables x, w, y, z, and h are binary. Moreover, the decision variable e belongs to the set of real numbers.

$$x, w, y, z, h \in \{0, 1\}, \ e \in R \quad (13)$$

4. Methodologies applied

This section presents the methods that have been applied in this work. The Generate and Solve method, described in [2], controls the application of genetic algorithm and Partition heuristics. Partition heuristics, in turn, is applied during the evaluation process of individuals of the genetic algorithm.

4.1. The Generate and Solve Methodology

Exact algorithms and metaheuristics are distinct, approaches to solving combinatorial optimization problems efficiently, each presenting advantages and disadvantages. Some of these hybrid algorithms aim to obtain optimal solutions with smaller execution times, while others seek to achieve better heuristic solutions [18]. The solutions of the instances generated by metaheuristics are determined when the subproblems are solved by an exact solver, which functions as a decoder. To achieve better heuristic solutions, the proposed hybrid methodology [19] suggests the integration of an accurate method to a metaheuristic generating small instances of the problem addressed, according to **Figure 1**. The hybrid mechanism is intended to determine a subproblem that can be solved correctly and provide a good solution to the original problem. The reductions made to the problem seek to limit the number of possible

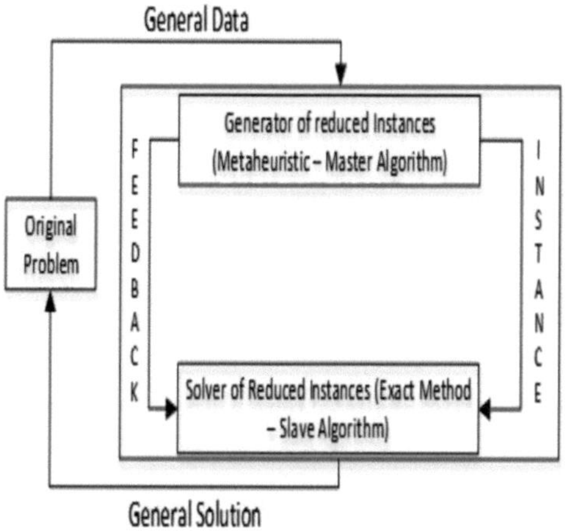

Figure 1. The hybrid framework.

solutions, always respecting all the restrictions imposed to the initial problem. In this way, the optimal solution found for each generated instance is also a viable solution for the problem to be solved. Also, the value of the objective function for the optimal solution of each subproblem, obtained from the exact method (encapsulated in the Solved of Reduced Instances (SRI) component), defines the quality of the generated instance. As shown in **Figure 1**, the information can be used to guide the search process of Generator of Reduced Instances (GRI).

4.2. Density control

When the genetic algorithm is applied to generate and solve methodology may occur a problem known as explosion density. To the extent that the generations are evolved, there is the possibility that the algorithm converges to a great place where the number of sensors participating chromosomes increases significantly, causing an alignment of subproblems of the original problem undermining the execution in time feasible and proper implementation of the methodology. To prevent this issue was introduced by [10] a density control operator. This operator is applied while generating reduced levels, after executing the intersection operator, replacing the mutation operator, as shown in **Figure 2**. The more disabling this sensor impair the value of the objective function, checking aspects of coverage, connectivity, and the penalty for not covering demand points, the more chances it must be maintained. This amount is credited only to chromosomes that are active after crossing operation. Inactive genes, i.e., equal to "0" did not receive credit.

Its purpose is to measure the density of the chromosomes; it is necessary to use a parameter called density ideal (DI). When the method is executed, a check of each chromosome density is made by comparing the credit value with value density ideal. If the density is less than or equal to DI, the chromosome remains unchanged. Otherwise, the genes with zero credits values that are active are disabled randomly. If the chromosome density is still higher than the DI even disabling genes with a null value, genes with lesser value become disabled until the density meets the density parameter ideal. In the latter case, there is a possibility of damaging the final solution, because genes that have some value for the objective function are disabled. Thus, this last step becomes optional and should be activated according to the application scenario. If time is less relevant, it is interesting to keep this step.

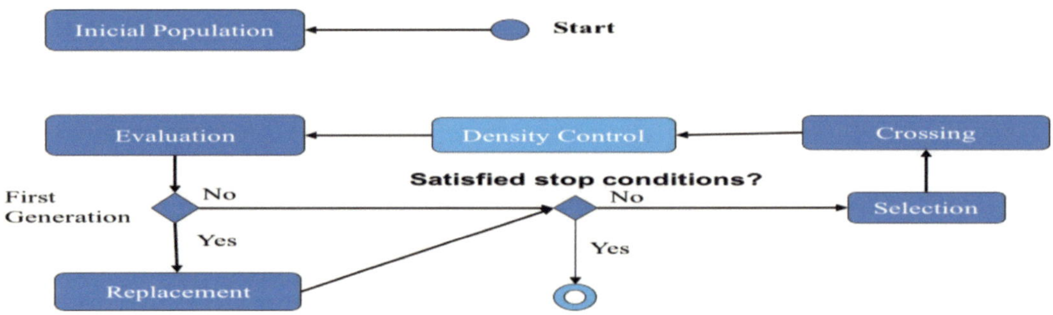

Figure 2. Genetic algorithm with density control.

4.3. Random key genetic algorithm (RKGA)

The genetic algorithm (GA) is chosen to be the Generator of Reduced Instances. To adapt the problem so the GA can be used, a reducible structure must be set maintaining the original's problem characteristics. Moreover, the main goal of this question is to indicate which sensors will be active in which time interval. The reducible structure matrix describes if a sensor can be activated or not in each time interval. **Table 1** presents the alleles that compose a chromosome for a network instance with three-time intervals and four sensors.

The genetic algorithms are applied in many solutions in problem classes. However, accordingly, to [1], the basic Genetic Algorithm has a difficult time to maintain feasible solutions when generating the offspring chromosomes. So, this might reduce the final result's quality. To mitigate this problem introduced a new genetic algorithm class called random key genetic algorithm. Those keys belong to real numbers set ($n \in \mathbb{R}$). Additionally, store the values; an array is used. A decoder, frequently a deterministic method, uses the random-key array as input data to return a problem's feasible solution. As the RKGA is an evolutionary process, it evolves a chromosome population, randomly generating the alleles into [0, 1] interval. After the decoder calculates the fitness value for each individual, the population is partitioned into two individual groups, the elite ones, and the non-elite ones. The first one, with fewer individuals, with higher fitness value, and the other ones, a bigger group, with smaller fitness value. Typically, the RKGA uses the elitism to create the next generation. Thus, all elite individuals are cloned from k generation to k+1. Following this, a mutation operator must be applied. The mutants are created using a random-key array like happens with the first generation. After the mutation, it is obtained a set of P − Pe − Pm, where P is the current population, Pe is the set of elite individuals, and Pm is the set of mutant individuals. Finally, the next generation is created using the elite, non-elite, and mutated individuals, generating the necessary number of individuals to reach the population's size. The crossover operation is executed by an operator introduced by [2], called parameterized uniform crossover. Given to parents A and B, stochastically selected from the current population, the child chromosome's alleles are chosen from parent A or B, respecting a probability P, for choose from one of them. **Figure 3** represents the RKGA generation k to generation k+1.

The application was proposed in two phases. The first step consists of constructing stables blocks and boxes and a second one that installs these blocks on the container floor. Conducted by Generate and Solve as its Reduced Instances Decoder. The experiments were made using library test cases, and the technique exceeded the best methodologies found in the literature for several cases. The decoding is made in three phases. First, a packing sequence of the type of the item is created, using a random-key array. Additionally, a rotation variance array is generated, and finally, it is decoded an allocation procedure array. Promising results are applying the RKGA comparing with approaches to solving the problem.

Array index	0	1	2	3	4	5	6	7	8	9	10	11
Alleles	1	0	0	1	0	1	1	0	0	1	0	1

Table 1. A chromosome alleles representing sensors by time intervals.

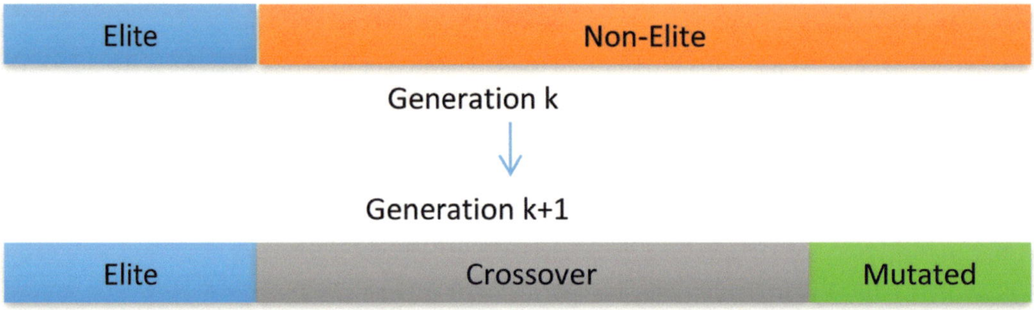

Figure 3. RGKA generation k to k+1.

4.4. Biased random-key genetic algorithm (BRKGA)

The main difference of the BRKGA to RGKA is that BRKGA has different ways to select parents to create the offspring during the crossover operation. In this one, the key array is composed of various genes, which are coded into the real number interval [0, 1]. Following this, as the RKGA and the conventional GA, a deterministic method is used to calculate the chromosomes fitness values and executing the remaining operations and generations until a stop condition is satisfied. In this work, three different decoders were used. First, it was used a decoder with job allocation's priority equals to the gene's value. A second decoder, where the combination of gene's value and ideal priority value give the priority. Finally, a third decoder is used. This one is a hybrid decoder that combines the other two, obtaining two solutions for one chromosome. It was concluded that the hybrid decoding method is appropriate to be applied to BRKGA to solve multi-objective problems, which many feasible solutions are necessary. Basically, for each crossover operation, it is generated a new array of random keys. The keys generated work as a probability value P_i. When executing the crossover operation, two random parents are chosen and, for each allele of the child chromosome, the parent to heir the gene is chosen accordingly to the random-key array. Each parent has a 50% probability to be selected for each gene. **Table 2** represents the process described.

4.5. Division of observation period

They are presented by [16] many of integer linear programming to optimize energy consumption but do not consider the use of dynamic schedule. The solution proposed by [1] is to create

Rk-array	0.58	0.42	0.11	0.83	0.38	0.75	0.68	0.39	0.21	0.94	0.46	0.88
Parent A	0	1	1	0	1	1	1	0	0	0	1	0
Child	0	1	0	0	1	1	1	0	0	0	1	0
Parent B	1	1	0	0	1	0	1	0	0	1	1	1

Table 2. BRGKA child generation accordingly to a random-key array.

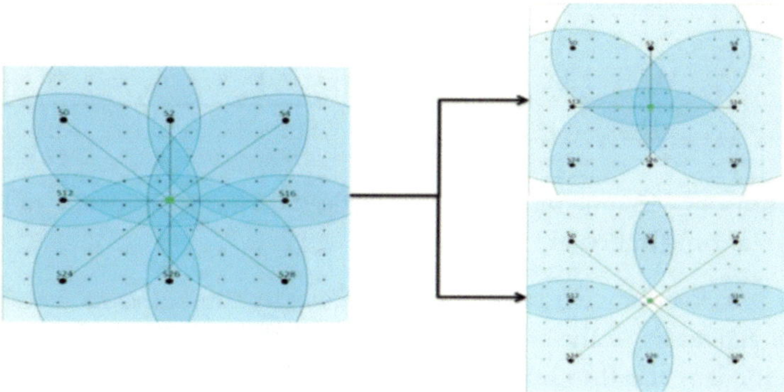

Figure 4. Division of the observation period.

different time intervals, which only activate the minimum number of sensors needed to cover all points of demand and to satisfy the connectivity restrictions. The use of different time intervals prevents premature depletion of some sensors, thus bringing a more homogenous level of consumption of the battery power of sensor nodes of the network. Considering only a distance of coverage for all monitored phenomena implies that the radius used shall correspond to the phenomenon that has the smallest radius of coverage. In the same way, just take a sample rate for all phenomena implies that the rate used should be the phenomenon that varies more often. Furthermore, in **Figure 4**, it corresponds to the left to the deployment of a network without using multiple time intervals, i.e., all the sensors are activated and are monitoring the region corresponding to its coverage radius to drain the battery power. So, this creates a large volume of redundant data, which should be treated not to harm analysis. In this case, when a sensor node is activated, it must be active to drain the battery power. Also, in **Figure 4** on the right shows the application's network division into periods, activating only the required nodes sensors to cover the maximum possible demand points. In this example, the lifetime of the network is doubled.

4.6. Implementation of genetic algorithm in WSN

Apply efficiently to Generate and Solve Methodology; it is necessary to find a way to reduce the problem size and represent it in a chromosome. Implementation of WSN applied in this study, the values of the genes indicate that the sensors in time and similar phenomena participate in the sub-problem. Therefore, the implementation of the methodology shows what restrictions or variables must be taken from the mathematical model according to the indications of the chromosome. The data structure used to represent the genes is a vector of binary numbers {0, 1}, where "0" indicates that the sensor node cannot be activated in the corresponding time interval and the value "1" indicates that a sensor node can be enabled in the relevant range. Many vector indexes are corresponding to the number of sensor nodes multiplied by the number of time intervals. **Table 3** shows a chromosome relating to an instance with three times intervals and four sensors. The indices 0 through 3 correspond to

0	1	2	3	4	5	6	7	8	9	10	11
0	1	1	0	1	0	0	1	1	0	1	0

Table 3. Chromosome, for instance, WSN homogeneous.

the genes about sensors 4 in time interval 1. Similarly, there is an indication of the values of the genes in subsequent time intervals.

The second way is to keep the chromosomes as is done in homogeneous networks, indicating the possibility of activation of the sensor at time intervals. The selection of individuals for the crossover operation is made through a technique called roulette. The chances that an individual should be selected corresponds to a share of proportional roulette to their fitness value. The fitness values of all individuals are joined in the total amount of roulette. Each individual has an equal share to its fitness value. It then generates a random number within the range comprising the total value of roulette. This value indicates which individual was selected. The crossover operation applied this GA was the Crossover one point. For each pair chosen to participate in the operation, is selected at random a cut-off point. The genetic material from each parent is divided and exchanged with each other, creating two new chromosomes. Each individual who will participate in the next generation undergoes the mutation procedure. Each chromosome gene has a probability of 0.5% of having its value altered. A random value between 0 and 1 is generated. If this value is less than 0.05, the allele undergoes the mutation.

4.7. Application of BRKGA Wireless Sensor Network Problems

Among the differences between the GA and BRKGA, one is the intersection operation. In execution the crossing of individuals applying the random key algorithm, a key vector is generated, and the operation is performed structured on the values of this vector. First, it was implemented RKGA to carry out the crossover operation. Through a standard acceptance value of 0.5, for example. It is decided whether the sensor will participate in the reduced problem instance. If the value of the random key is greater than the acceptance value, then the sensor part of the reduced instance, as exemplified in **Table 4**.

Enhance application of the algorithm in WSN problem; the BRKGA algorithm was modified. This time, instead of indicating whether a sensor may be activated at a time interval only if the value of a key is greater than the acceptance value, the algorithm generates, for each allele a random value and compares it with the corresponding key. Thus, the higher the value of your key, the more chances you have a sensor to participate in the sub-problem. Thus, even if one allele has a random key with a little value. It can still be selected to participate in the reduced instance, even if their chances are lower. **Table 5** demonstrates the application.

4.8. Heuristic Partition

The application Generate and Solve Methodology significantly reduces the original problem into smaller problems maintaining its characteristics, allowing solutions to be found next to the solutions of the original problem. However, even with the improvement obtained by the

Index	0	1	2	3	4	5	6	7	8	9	10	11	12	13	14	15	16	17	18	19	20	21	22	23
Gene	0	1	1	0	0	1	1	1	0	0	1	1	1	0	1	0	1	1	1	1	0	0	0	1
Key	0.32	0.73	0.54	0.28	0.13	0.84	0.56	0.73	0.25	0.41	0.55	0.68	0.56	0.18	0.92	0.37	0.76	0.51	0.65	0.85	0.08	0.36	0.22	0.89

Table 4. Random keys BRKGA.

Index	0	1	2	3	4	5	6	7	8	9	10	11	12	13	14	15	16	17	18	19	20	21	22	23
Gene	0	1	1	0	0	1	1	1	0	0	1	1	1	0	1	0	1	1	1	1	0	0	0	1
Key	0.72	0.3	0.44	0.28	0.13	0.84	0.56	0.73	0.65	0.41	0.55	0.68	0.56	0.18	0.92	0.37	0.36	0.48	0.65	0.85	0.08	0.36	0.22	0.89

Table 5. Random keys adapted BRKGA.

methodology are instances where even higher, the runtime ends up again impractical. The Generate and Solve Methodology is the reduction of the original problem into subproblems, less complex, preserving the original features of the problem. Similarly, the Heuristic Partitioning is also to reduce the size of the problems. It subdivides the sensing region of the wireless network into smaller pieces, creating subnets of an original network, generating partitions, the Generate and Solve Methodology should be applied. Additionally, variables that do not belong to those partitions are discarded during the execution of each party. This fact reduces the complexity of the problem. After the solution of each partition, it must perform the junction of the results of each partition to get the results for the original problem. This heuristic must have low run-time, as a real solution had been obtained previously. The larger the sensor the coverage radius for a phenomenon, fewer sensors are required to cover the entire region of interest. So, this should be considered to calculate the optimal size of the partitions. When partitions are generated based on the phenomenon that has a greater distance to cover, the size of them may end up too big, getting very close to the size of the original network. Therefore, the time to solve large problem instances will not be feasible. However, if the partition sizes are too small, when the junction of the partitions is done there will be many active sensors in a small area, especially when there are phenomena whose coverage rays are large.

4.8.1. Application forms

Two ways of application of Partition heuristics were tested. In the first case, the wireless network is divided into partitions. Thus, each partition is treated in isolation or is regarded as a different network and the Generate and Solve then executed one time for each partition generated. Thus, when the chromosomes of the genetic algorithm are generated, some alleles correspond only to the number of variables relating to the partition that is being resolved. **Figure 5** is the problem of the execution flow by applying this way:

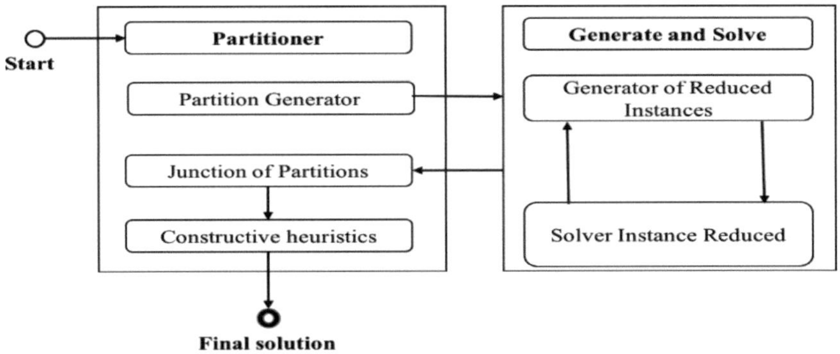

Figure 5. Partition Heuristic flow applied before Generate and Solve.

This approach is more efficient in problems in that all phenomena network coverage rays are identical. The size of the partitions is too large, it is very close to the original problem, making it impractical. It was then implemented another form of application of Partition heuristics, to solve this issue. This time, it is executed within the Generate and Solve Methodology. When the genetic algorithm performs the assessment of the chromosome, it runs the exact method to get the result of the sub-problem. Only this time the partitions are generated for the problem. So, every individual assessed, the exact method is executed n times, where n is the number of partitions. In an other manner, the chromosomes generated for each individual had the size corresponding to the variables of the corresponding partition. In this case, the treated chromosomes are the same size as the original problem, but each partition only considers the values of the genes found on chromosomes that match your partition. The key to this approach is precisely at this point. For after the junction of the partition, the value obtained will be the result of the chromosome in question. Then, in the case of phenomena with a large radius of coverage, the result of the separate partition may contain time intervals in which there is no active sensor. For the separate partition that corresponds to a bad result, injuring coverage restriction. However, there are high chances of a sensor on another partition cover the region due to its large radius coverage. So, **Figure 6** shows the flow for implementing.

Another key point to obtain a good solution implement the second approach is to facilitate the partitions exist solutions without active sensor at some time intervals. At this point, density control operator has fundamental importance. It will generate instances where there is no possibility of any sensor cover a region of the monitored space. For this, the operator must be applied separately to chromosome regions representing each phenomenon, on a chromosome where allele 1–8, the values correspond to the temperature phenomenon, which has smaller coverage radius, and allele 9–16. The values correspond to the allele light phenomenon, which has a larger radius of coverage. After solving the problems related to the partitions, it must perform the junction of the partitions back to the original state. At this point, it is necessary to calculate the net energy consumption. The power consumption of each sensor of each partition should be added since no sensors were sharing between partitions. The calculation of the penalty for not demand coverage points should be made after the joint because in some cases, demand points that were not covered by a sensor of a partition can be covered by other sensors which were in another partition. In some cases, sensors can be activated very close to each other between different partitions when they are close to the circumferences of the partitions. A constructive heuristic should be executed to perform the junction of the partitions to optimize this problem. This heuristic adds a new layer in wireless sensor networks problem by improving the performance of the implementation of the problem, trying to solve it in time

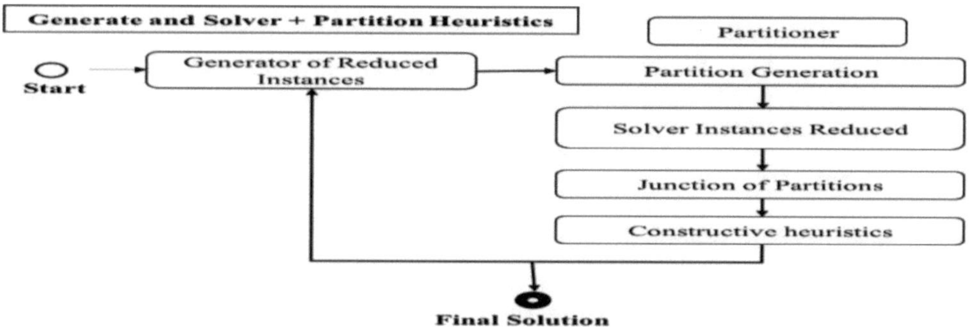

Figure 6. The flow of Partition Heuristics applied internally in Generate and Solve.

viable, keeping the restrictions of coverage and connectivity. The method selected to apply the WSN problem was nearest neighbor, due to its low complexity and the initial solution is the best solution that the methodology applied encountered so far. Neighbors were generated according to the possibility of sensor activation according to the remaining power of their respective batteries following the following rules:

- Sensor activation combinations are generated that still have enough energy at different time intervals. In some cases, the result obtained by the hybrid method can end up leaving the region of a sub-area of interest with the possibility of finding a sensor cover it. So, neighbors are generated starting from this premise, where sensors still have energy are activated and evaluated. This rule holds just one activation neighbor. That is, if more sensors can be activated either at different time intervals or the same interval, only one will be enabled. The remaining sensors will generate new neighbors. In such cases, the activation of more than one sensor on the first individual is performed in subsequent generations.
- It is made of a sensor off when there are two active sensors very close to each other in the same time interval. Sensors monitor tend to close redundant areas, consuming energy that could be used in other time intervals. Disabling these sensors involves the reduction of the objective function value and, as the goal is to minimize the total consumption of the network, the neighbors generated with this rule have a higher value of fitness. This rule is checked whether the candidate sensor to deactivate the route is part of another sensor in the same time interval. To recalculate the objective function, it needs to check in what point interval each sensor is active and which phenomena each is monitoring. Thus, it is possible to identify how much energy each sensor starts to consume according to data consumption, MOTE Battery Life Calculator [17].

5. Computational results

The intentionally of the model is to optimize the energy consumption without homogeneous wireless sensor networks. The number of time intervals has fundamental importance for this period of the total time of network life is obtained by multiplying the number of intervals for the duration of each of them. However, networks of wireless sensors are usually allocated in severe environments, so it is necessary to extend the maximum time that the network is active without compromising its reliability. The combination of the number of time intervals, sensors, demand points, phenomena, an array of sensors and types of genetic algorithms make different scenarios. Since the purpose of the model is to optimize the lifetime of the network, have been generated scenarios first structured in the number of time intervals of 2, 4, 6, 8, 10, 12, 14, 16, 18 and 20 times intervals. For each of these scenarios, instances were generated structured in a combination of features: Sensor Nodes Quantity: 16 sensors, Phenomena: Temperature, Sensor allocation: in a grid or randomly and Genetic Algorithm Type: AG or BRKGA. The instances with 16 sensors nodes were created, and to instances of 16 sensors were allocated, 100 points of demand. In turn, for each one of these bodies were set other sub-sets with the sensors arranged in grid form and randomly arranged sensors. Then, for each of these subsets, others were performed by applying the AG and the other with BRKGA. The values shown in the tables is the arithmetic mean of the values found in 10 executions. The temperature phenomenon was used in scenarios with just a phenomenon, with an 8.8 m coverage radius. The demand points were arranged in a regular grid in all scenarios, in an area of 20 × 20 m with a demand point per m^2. Each demand point is associated with a phenomenon. Then a demand point can be served to a phenomenon and not the other, in the same time interval. The sensor nodes when distributed in grid form, are allocated analogously to demand points. Otherwise, the position of each sensor is chosen at random using a uniform distribution function for generating the network. The sensor transmission range, which allows the exchange of information between them to arrive at a sink node, was 11 m. All sensor nodes have the same characteristics of transmission, reception, coverage, connectivity, and battery capacity. Only one sink node has been allocated to the centers of the monitoring area in all instances. All network elements were generated by applying concepts of geographic coordinates. The power consumption values were calculated based on the values found in the sensor manual. The total transmission and reception relating to monitoring data have been computed based on the amount of data transmitted by the devices. It was awarded a penalty considerably high value not to allow the model disable sensors unnecessarily increasing the number of demand points not covered. The results are displayed actual values with the values of the penalty and the values representing the energy used by the sensors. Moreover, the model of integer linear programming was implemented in the Java programming language 7, using the CPLEX library 12.1 academic licenses, responsible for integrating with the solver used. The computer used has an Intel Core i7 fourth generation processor, 8GB of RAM, running Windows 7 operating system 64-bit. It is emphasized that the value of the objective function is the sum of the energy used by all sensor nodes during all time intervals with the values of the penalties applied when a demand point is not covered. However, for practical purposes, the value that is only the net energy expenditure is also important. Thus, tables shows the total value as "Value Obj" and the value representing only the energy used by the sensors as "Energy." The "% not covered" represents the average percentage of demand points not covered. To emphasize the need to apply heuristics to the problem, **Table 6** shows the results obtained by performing only the exact method for scenarios with 16 sensors, both cases one allocation grid sensors and random. The time required to carry out the scenarios is impractical. So, it is because the only difference between the instances is some sensor nodes.

It is a remarkable increase in the size of cases 100% about some time intervals. However, in cases with a phenomenon, 12, 14 and 16 times intervals, it is remarkable that the value of the objective function remains with similar values across instances, and some demand points not

16 Sensors/1 phenomenon								
Time intervals	Variables	Constraints	Grid			Random		
			Value Obj	% not covered	Time	Value Obj	% not covered	Time
2	12217.00	19648.00	5376.22	0.00	0.03	5376.22	0.00	0.03
4	48689.00	78384.00	10752.43	0.00	0.08	13203.82	0.02	0.12
6	73025.00	117552.00	15698.30	0.00	0.12	15698.30	0.04	0.13
8	97361.00	156720.00	21074.52	0.01	1.97	21358.16	0.04	25.18

Table 6. Execution with ILP, 16 sensor nodes.

covered increases dramatically. Moreover, this is due to the total collapse of the network energy. However, as in such cases, the energy used tends to be higher, depletion of energy is already noticeable on executions with eight times intervals on, therefore, is not necessary to run instances with larger amounts of time intervals. In such cases, the number of demand points increases according to the number of time intervals, since there would be no more sensors to meet the request. It is possible to observe the instances 12, 14 and 16 times intervals implemented with BRKGA that the energy expended values are identical in **Table 7**. So, this was due to the depletion of the energy of all the receivers.

1 Phenomenon/16 sensors/allocation grid								
Time intervals	**Partition + GA**				**Partition + BRKGA**			
	Value Obj	**Energy**	**Not covered**	**Time**	**Value Obj**	**Energy**	**Not covered**	**Time**
2	5376.22	5376.22	0.00	1.41	5376.22	5376.22	0.00	1.33
4	15831.43	10752.43	2.75	2.15	14256.43	10752.43	2.92	1.61
6	23078.30	15698.30	3.50	3.09	22028.30	15698.30	3.52	2.60
8	30800.52	21074.52	4.05	7.77	30692.52	21074.52	4.01	4.51
10	40991.74	26450.74	4.85	6.58	40910.74	26450.74	4.82	4.70
12	68753.18	29483.18	10.91	6.33	54565.61	31396.61	6.44	4.59
14	164963.53	30278.53	32.07	7.89	130714.61	31396.61	23.65	5.50
16	226349.95	31826.95	40.53	9.30	173134.61	31396.61	29.53	6.17
1 Phenomenon/16 sensors/allocation random								
2	8273.52	8273.52	0.00	1.95	8273.52	8273.52	0.00	1.06
4	14802.82	13203.82	0.00	2.49	14913.82	13203.82	0.00	1.68
6	23633.30	15698.30	3.50	3.63	23744.30	15698.30	3.50	3.07
8	55332.23	23637.23	13.21	6.24	54979.36	23242.36	12.86	5.45
10	84676.68	23938.68	20.25	7.74	83176.50	24664.50	19.50	7.47
12	131714.63	23762.63	29.99	7.91	114270.74	25509.74	24.66	7.46
14	179129.22	27128.22	36.19	9.64	178187.74	26150.74	34.87	6.87
16	228660.20	27549.20	41.90	14.79	216423.20	27687.20	39.32	10.18

Table 7. One phenomenon, 16 sensors, and 100 demand points.

6. Conclusions and future work

Despite the growing advancement of the processing power of today's hardware, it is not possible to perform exact methods to solve the problem of coverage and connectivity in Wireless Sensor Networks. The Generate and Solve Methodology was implemented to solve the problem in homogeneous networks [10]. However, as the network grows instances, the execution time tends to be impractical again. The Heuristic Partitioning layer adds a further problem, further reducing the time required to find a feasible solution with good instances 100% higher in both types of networks. Not only it is remarkable the possibility of growth scenarios, but also the quality of the results, in turn, exceeded the results found in the literature on total consumption of network energy and demand points coverage. There are scenarios in which some sensor nodes are far away from the sink node. Thus, they cannot communicate directly and require others node (s) sensor (s) to provide the route to the sink. However, particularly in instances with many time intervals, there is the tendency of all the sensors that could promote this route does not possess sufficient energy to do so. In such cases, these sensors end up being little use or even not used, in general, leaving demand points not covered. One possibility to solve this problem is to allocate some extra sensors to the network. The weakness of this method is that the inclusion of sensors leads to increased complexity, which should not significantly increase the runtime by applying the methodology set. It has been seen that with the applied genetic algorithms solutions can be found in less time than with BRKGA. Tests were performed with a larger battery capacity of sensor nodes to test how far it can increase the complexity of the scenarios solving them infeasible time. A significant increase in the number of time intervals was observed. In some situations, it was found an increase of 300% compared with the amount of execution time intervals in the literature. Therefore the ability of the methodology to be even closer to real network instances; they are allocated some sensor nodes greater than the amount ranges of the tested scenarios. Also, used results found in [10] for comparison. In it, the coding of chromosomes occurs differently, using a fixed number of alleles. These alleles are represented indices of sensors participating in the problem.

As future work, we highlight the application of coding the problem of heterogeneous wireless sensor networks and as well as the Partition Heuristic. Another future work would consider Heuristic partition in the container loading problem in parallel to Generate and Solve Methodology which was initially, applied to this problem in [2, 13]. The challenge lies in how to produce partitions without hurting the quality of results and maintains the possibility of allocating all kinds of items required within the container.

Acknowledgements

The authors are thankful to National Counsel of Technological and Scientific Development (CNPq), Foundation for Support of Scientific and Technological Development Ceara State (FUNCAP) and Coordination for the Improvement of Higher Level or Education-Personnel (CAPES) for the support received on this project.

Author details

Plácido Rogerio Pinheiro*, Álvaro Meneses Sobreira Neto, Alexei Barbosa Aguiar and Pedro Gabriel Calíope Dantas Pinheiro

*Address all correspondence to: placidrp@uol.com.br

University of Fortaleza – UNIFOR, Graduate Program in Applied Informatics, Fortaleza, CE, Brazil

References

[1] Aguiar AB. Tackling the problem of dynamic coverage and connectivity in wireless sensor networks with an extended version of the generate and solve methodology generate and solve methodology [Master's dissertation]. Graduate Program in Applied Informatics, University of Fortaleza, Brazil; 2009

[2] Aguiar A., Pinheiro PR, Coelho ALV, Nepomuceno N, Neto A, Cunha RPP. Scalability analysis of a novel integer programming model to deal with energy consumption in heterogeneous wireless sensor networks. Communications in Computer and Information Science. 2008;14:11-20

[3] Aguiar AB, Pinheiro PR, Coelho ALV. On the concept of density control and its application to a hybrid optimization framework: An investigation into cutting problems. Computers & Industrial Engineering. 2011;61:463-472

[4] Aguiar AB, Pinheiro PR, Coelho ALV. Optimizing energy consumption in heterogeneous wireless sensor networks: A novel integer programming model. In: Proceedings of the IV International Conference on Operational Research for Development – ICORD; 2007. pp. 496-505

[5] Amaro Junior B, Pinheiro PR, Saraiva RD, Pinheiro PGCD. Handing the random-key genetic algorithm for solving the nesting problem. In: Anais do XLVI Simpósio Brasileiro de Pesquisa Operacional; 2014

[6] Blum C, Roli A. Metaheuristics in combinatorial optimization: Overview and conceptual comparison. ACM Computing Surveys. 2003;3:268-308

[7] Eiben AE, Smith JE. Introduction to Evolutionary Computing. Springer; Berlin, Heidelberg, 2003

[8] Quintão FP, Nakamura FG, Mateus GR. Evolutionary algorithms for the dynamic coverage problem applied to wireless sensor networks design. Proceedings of the IEEE Congress on Evolutionary Computation. 2005;2:1589-1596

[9] Talbi G. A taxonomy of hybrid metaheuristics. Journal of Heuristics. 2002;8:541-564.

[10] Harold Ishebabi, Philipp Mahr, Christophe Bobda, Martin Gebser, and Torsten Schaub, "Answer Set versus Integer Linear Programming for Automatic Synthesis of Multiprocessor Systems from Real-Time Parallel Programs," International Journal of Reconfigurable Computing, vol. 2009, Article ID 863630, 11 pages, 2009.

[11] Zhou H, Liang T, Xu C, Xie J. Multiobjective coverage control strategy for energy-efficient wireless sensor networks. International Journal of Distributed Sensor Networks. vol. 2012, Article ID 720734, 10 pages, 2012

[12] Gonçalves JF, Almeida J. A hybrid genetic algorithm for assembly line balancing. Journal of Heuristics. 2002;8:629-642

[13] Puchinger J, Raidl G. Combining metaheuristics and exact algorithms in combinatorial optimization: A survey and classification. In: Proceedings of the 1st International Work-Conference on the Interplay Between Natural and Artificial Computation, Lecture Notes in Computer Science. Vol. 3562; 2005. pp. 41-53

[14] Nguyen K, Nguyen T, Cheung S-C. On reducing communication energy using cross-sensor coding technique. International Journal of Distributed Sensor Networks. 2011;7:12, Article ID 837128.

[15] Wolsey LA. Integer Programming. John Wiley & Sons; USA, 1998

[16] Dumitrescu L, Stützle T. Combinations of local search and exact algorithms. In: Applications of Evolutionary Computing, Eds: Stefano Cagnoni, Colin G. Johnson, Juan J. Romero Cardalda, Elena Marchiori, David W. Corne, Jean-Arcady Meyer, Jens Gottlieb, Martin Middendorf, Agnès Guillot, Günther R. Raidl, Emma Hart, Lecture Notes in Computer Science. Vol. 2611; 2003. pp. 211-2247

[17] MOTE Battery Life Calculator [Internet]. May 2007. Available in: <http://www.xbow.com/Support/Sypport_pdf_files/PowerManagement.xls>

[18] Pinheiro PR, Coelho ALV, Sobreira Neto AM, Aguiar AB. Towards aid by generate and solve methodology: Application to the problem of coverage and connectivity in wireless sensor networks. International Journal of Distributed Sensor Networks. 2012;8:1-11.

[19] Pinheiro PR, Sobreira Neto AM, Aguiar AB. Handing optimization energy consumption in heterogeneous wireless sensor networks. International Journal of Distributed Sensor Networks. 2013;2013:1-9

Section 2

WSN Implementations

Chapter 5

WSN in Conservation Management

Akbar Ghobakhlou and Shane Inder

Additional information is available at the end of the chapter

http://dx.doi.org/10.5772/intechopen.70383

Abstract

A wireless sensor network (WSN) utilising a mesh configuration is a cost-effective and labour-saving solution for remotely monitoring traps and tracking devices used in conservation management. The unintentional introduction of stoats and rats into a once pristine ecosystem has resulted in the devastation of large parts of New Zealand's native flora and fauna. Other equally harmful mammalian species, including possum, for their fur, and domestic cats, were introduced intentionally. Abundant vegetation and a lack of predators lead to rampant population growth, further exacerbating their destructive impact. Effective monitoring, trapping and control of mammalian pests have proven difficult, time-consuming and expensive, primarily relying on socially controversial methods such as aerially delivered toxins. Despite advances in technology, costly and time-intensive manual checking of lures, toxins, traps and tracking devices remains a limiting factor. Together with WSN-based remote monitoring capability, these advances look set to have a significant impact. This chapter discusses opportunities for WSN in conservation management. It outlines a mammalian pest management project utilising a series of possum-specific self-resetting traps. A WSN designed for remotely monitoring possum trap activity is detailed, and the process for reconfiguring and presenting field-trial data via alpha-numeric and graphical user interface applications is described.

Keywords: WSN, mesh network, pest management, possum control, remote monitoring

1. Introduction

Introduced invertebrate pests, such as rats, mustelids (ferrets; Mustela furo and stoats; Mustela erminea) and possums (richosurus Vulpecula), have and continue to cause irreparable damage to New Zealand's flora and fauna. The greatest impact can be seen in New Zealand's avifauna where 42% of land bird species abundant prior to human habitation are now extinct. Of the remaining 155 native species, 45 (29%) are now endangered [1]. As the primary maintenance,

INTECH
open science | open minds

host of bovine tuberculosis (bTB), possum in particular, pose not only an ecological threat but also a significant economic one to New Zealand. The main goal of TBFree New Zealand (formerly Animal Health Board) is to manage intensive control of brushtail possum as the principal vector of bTB. TBFree NZ funded research has supported the development of cost-effective, environmentally acceptable tools and tactics for wildlife disease surveillance and control [2]. Fundamental to the success of TBFree's eradication strategy is the role of research in developing and testing innovative improvements to cost-effective control and surveillance technologies. Researchers are gaining a better understanding of the behaviours of possums at low densities and developing a suite of detection, possum-specific toxins and application methods and pest control systems, aimed at sustainable pest management [3]. Significant efforts are being made to create mammalian pest-free islands and sanctuaries equipped with predator-free fences according to a 2014 Royal Society of New Zealand 'Expert Advice Paper'. To avoid reinvasion, the paper continues to describe the need for humane methods of eliminating pests at landscape scale. Along with greater public support, it also suggests maintenance, involving effective large-scale monitoring, identification of pest species and trapping of reinvaders.

In order to address some of these issues, a number of interdisciplinary teams of researchers from across New Zealand have engaged to develop new tools and techniques for conservation management. Working across multiple projects, the teams have developed species detection and monitoring devices that incorporate advanced digital technologies; a suite of advanced humane toxins; targeted, electromechanical toxin delivery devices and systems and long-lasting baits and lure systems. Academic teams were linked to industry partners to commercialise viable outputs and disseminate best practice approaches. Teams included ecologists, toxicologists, engineers, animal behaviour experts and industrial designers. Working in interdisciplinary project teams has generated benefits for all involved. Designers, for example gained expert knowledge, access to innovative ecological and technological developments and to testing facilities that were utilised in the design process, ensuring outcomes were fit for purpose and maintained longevity throughout the validation process and into the marketplace. Experts in conservation science benefited from the engagement by seeing the transformation of data from their research into tangible and readily influential form for iterative development and into commercialisation. As identified by Root-Bernstein and Ladle [4], conservation scientists do not commonly have the expertise to turn their data into functional products.

The user-centred approach was utilised to investigate and evaluate not only the technical requirements but also stakeholder and user needs to help contextualise the projects.

In order to gain an empathetic understanding of user needs, limitations and context, qualitative research methods were used. These included stakeholder interviews, objective observation and photo-ethnography during field trips with a diverse group of conservation staff, engineers and designers. Functional and user-focussed constraints informing the development of prototypes drew from data gathered. Field trips also gave first-hand insights into the complexities of ecological systems and the interactions and behaviours of the organisms within. Opportunities for product or system efficiency gains were highlighted by the difficulties faced by conservation workers and the labour intensiveness of existing processes such as data collection and device management [5, 6]. Qualitative research supplemented iterative prototyping and testing to inform continual development.

Although this research has delivered a suite of individual tools to use in the battle against mammalian pests, from a holistic perspective, there are a number of opportunities yet to be addressed. The time involved in set-up, checking, resupplying lures and resetting devices is still significant, particularly in some very large, often inaccessible areas of New Zealand's conservation estate. The ability to remotely access the data generated and stored in these digital devices would greatly enhance their efficiency as conservation management tools. Remote access to devices to monitor activity would reduce the frequency of manually checking devices, retrieving data and replenishing baits and lures.

Utilising a robust wireless sensor network (WSN) to remotely access and/or control devices may prove to be a feasible option. They are a relatively low cost, mature technology, which permits communication over a large area with a network of simple devices. Their use for detection and tracking purposes has already been demonstrated in diverse works [7–10]. Two projects have emerged that will demonstrate the efficiency gains possible with the integration of a WSN. One in the area of species recognition and monitoring and another a targeted pest management tool. It is anticipated that an ability to remotely monitor these tools will enable a more holistic, strategic and efficient conservation management approach.

2. Related work—potential applications in conservation management

Pest management and monitoring costs New Zealand's Department of Conservation around 10 percent of its total $450 million budget [10]. While advances in technology have seen the introduction of a suite of products aimed at efficiency gains, set-up and monitoring these devices remain a manual operation.

A significant amount of terrestrial wildlife management involves the use of landscape-scale tracking, trapping and/or pest eradication programs on government or conservation land tenures. These programs aimed at control and research on estimating species densities or investigation their ecology, which include live trapping programs to capture animals for study. Jones et al. [6] estimated that cost savings of up to 70% could be accrued from the use of WSN-enabled systems.

2.1. Monitoring

Analysing footprints on ink blotted paper, caused by animals walking through simple tracking tunnels, is a technique first described in 1977 [7]. The tunnels (**Figure 1**) are inexpensive and reasonably sensitive to the presence of rodents (particularly rats) when they are present at low densities less so with mustelids (weasels, ferrets and stoats) and possum. Wet weather and overtracking are key limitations of tracking tunnels, which can lead to data loss [11]. High population densities can lead to overtracking making identification of discrete footprints difficult. The paper or card substrate used to collect print data is prone to becoming illegible due to wet weather. Modifications to tunnels or more frequent paper changes can overcome this, however, at additional expense in equipment and time.

An interdisciplinary team, including ecologists, engineers and designers, from Lincoln University and Auckland University of Technology has developed an animal tracking system that addresses the shortcomings of traditional tracking tunnels. Using 'touch screen' technology, the Print Acquisition for Wildlife Surveillance (PAWS) device digitally records animal interactions. Time, date and weight data are collected along with the animals' footprint files. A field worker collects the digital data via a Bluetooth-enabled smartphone. Files are later uploaded to a computer for analysis and utilising a tailored algorithm, identify species with near perfect accuracy. Trials have proven that they are far more successful at monitoring pest animals than traditional options such as tracking tunnels, wax blocks or chew cards [12].

A PAWS device (**Figure 2**) utilising data transfer over a WSN would have significant advantages over manual data collection. Low power consumption, long-life batteries and/or solar recharging allow PAWS devices to remain active in the field indefinitely, continually relaying data. Reconfiguring the data and mapping species detection in real-time would allow conservation groups to utilise limited time and financial resources in areas that require immediate attention, such as pest management, and establish strategic programs to avoid reinfestation.

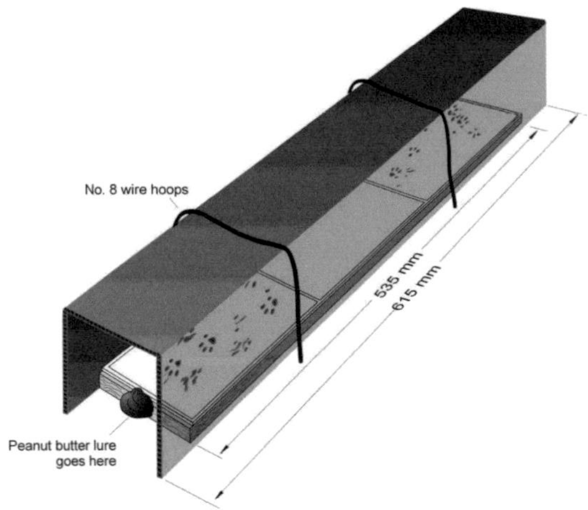

Figure 1. Traditional tracking tunnels used to monitor rodents and mustelids, DOC tracking tunnel guide v2.5.2, adopted from [7].

Figure 2. Print Acquisition for Wildlife Surveillance (PAWS) concept (adopted from [12]).

2.2. Pest management

Single kill traps and aerial distribution of 1080 are the primary means of controlling pest populations in New Zealand currently. Despite widespread public concerns over its use, the Environmental Risk Management Authority (ERMA) and Animal Health Board (AHB) determined in a 2007 assessment that Aerial application of 1080 toxin in cereal pallets, accounting for around 440,000 ha or 5% of conservation land [9] each year, has significant benefits for New Zealand's environment [8]. According to the report, 'These benefits would not be fully realised if the use of 1080 were restricted to ground-based operations only'. This is largely due to the inaccessibility of large parts of New Zealand's conservation estate and the high financial cost and labour intensity associated with ground-based pest control methods.

The NZ Department of Conservation describes the DOC200 box (see **Figure 3**) configured with a single-kill, spring-activated trap as 'best practice' for the control of rats, stoats and hedgehogs. This current paradigm for pest control has a proven track record in the field, is robust and is familiar to users. However, this approach is also time-consuming and expensive, requiring constant checking, resetting and lure/bait resupply.

The team was asked to deliver a reliable, cost-effective, safe and humane method for the extermination of large numbers of possums (100 toxicant doses per unit). In order to 'gain public support for mammalian pest control or eradication, especially where this involves toxins' [14], an interdisciplinary team was contracted to develop a possum focused targeted delivery system to exploit new humane toxins that had been developed. These toxins aimed to overcome environmental and economical inherent in aerially delivered or repeated manual replenishment of bait stations [13]. Species specificity, economical to produce and an ability to withstand being active in the field for over 12 months were also critical requirements of the system. It needed to be lightweight, low maintenance, robust and house the toxin securely [5].

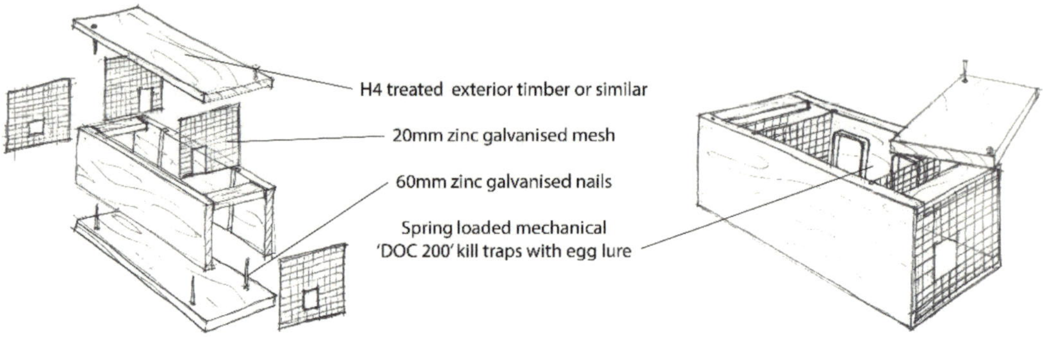

Figure 3. The DOC 200 approved humane kill trap for stoats, rats and hedgehogs [13].

A 2013 Field trial of prototype (see **Figure 4**) possum-specific toxin delivery units validated the efficacy of the toxin delivery devices. However, the trial proved to be very labour intensive due to daily frequent manual checking of traps, replenishing lures, downloading data logs and ensuring on-going functionality. These early trials utilised motion-detecting cameras at each device and pre-attached proximity collars on possums and devices (**Figure 4**). Reviewing radio-tracking signals from the 13 pre-collared possums indicated possum mortality and enabled the verification and collection of collars. **Table 1** shows station/mortality collar data downloaded for evaluation. Testing 10 of the 13 collared possums killed by spitfires with 8 of 9 of those collared in the site with a further 2 of 4 possums collared outside site. Uncollared possums seen after most collared had been killed. Most of the device activations occurred after the devices had been out for over 1 week. No non-targets were sprayed (mice, rabbits and pigs were present) indicating the triggering mechanisms worked as intended [12].

The field trail proved to be a successful approach to possum-specific pest control in this context; however, daily monitoring of the system in difficult terrain was expensive and labour intensive. It was therefore recommended that researchers explore methods to remotely access data and to decrease the necessity to check devices in the field.

Figure 4. Field trial devices configured to capture firing event data, including date, time, and location and snapshot evidence.

Device number	No. of activations	Observations
SF1	1	10 out of 13 collared possums killed by spitfires and 8 out of 9 of those collared in the site. 2 out of 4 possums collared outside site.
SF2	2	
SF3	3	
SF4	1	Most fires occurred one week after devices were installed on site.
SF5	1	
SF6	1	Un-collared possums seen after most collared gone and no non-targets sprayed; mice, rabbits were pigs present at site.
SF8	8	
SF9	4	Juveniles are not heavy enough to trigger the spitfires. Lighter trigger required for juveniles.
SF10	1	
SF11	1	Some possums removed/accessed bait by sitting on hood. Possums seen groom shortly after toxin application
SF12	5	

Table 1. Log of triggering events and field trial summary over 3-week trial period.

3. Wireless sensor network

A wireless sensor network (WSN) is a network formed by large number of spatially distributed autonomous devices known as nodes that use sensors to monitor physical or environmental conditions. These nodes are constrained by limited storage and power, similar to embedded systems. Numerous applications have been developed employing WSN technologies in many fields, including agricultural monitoring [7, 11] and animal behaviour studies [8, 13, 14]. **Figure 5** illustrates one of the commonly used WSN architecture in monitoring applications with sensor nodes, a sink node, a base station and a server. The sink node is the network coordinator to establish the network communication and send/receive nodes data within the WSN.

The radio frequencies used by communication modules in sensor networks are restricted by licensing. The selection of transmission frequencies is most influenced by their ability to transmit across undulating topography and/or through dense vegetation. As a general rule, the higher the frequency the more direct line of sight required [15]. The commonly used 2.4 GHz frequency has been adopted by most commercial network manufacturers since this and higher frequencies have a lower risk of data corruption when information is being transmitted.

3.1. WSN topologies

The development of WSNs has taken traditional network topologies in new directions. There are four basic WSN topologies for establishing a WSN as follows: peer-to-peer, star, tree and mesh. The data path between two nodes or a node and the gateway is referred to as a single-hop network. One of the most fundamental design choices is whether to use a single-hop or a multi-hop network. Multi-hop networks are useful in situations where measuring stations

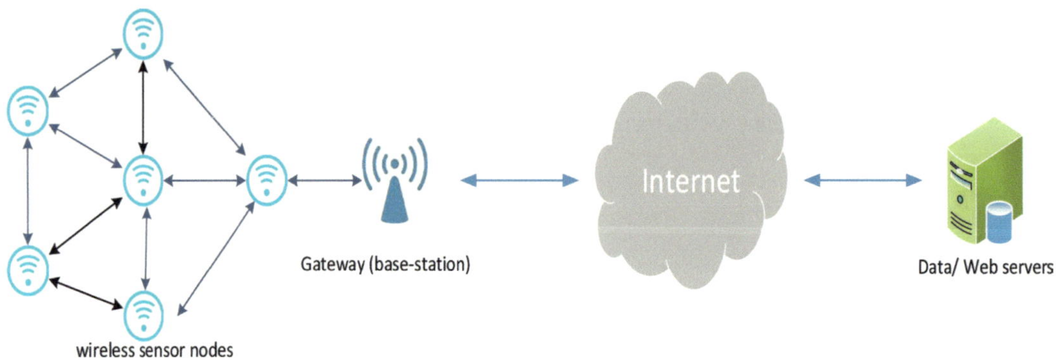

Figure 5. Schematic view of a wireless sensor network with mesh topology where data are collected through sensor nodes. All nodes communicate with each other and transmit data via a gateway or "base-station." The gateway transfers data to the data server using internet or satellite.

are not centred around a base station. Such scenarios may occur when monitoring a mountain side, a ridge, a lakeside or any other area that is elongated [16].

Mesh networks is a more complex network configuration where a node may needs to communicate with number of other nodes to make successful transmission. In order to extend the range of a network or avoid an obstacle, a wireless relay node can be added between a gateway and a leaf node. The mesh network consists of three types of node, namely sensor node, relay node and gateway node.

The sensor node is a device used for integration with the physical system that has been designed to monitor and/or control. The relay node is usually called 'routers' and used to extend network coverage area and provide back-up routes in case of network congestion or device failure. A sensor node may transmit data to the gateway node through multi-hopping. This is particularly important in areas where physical terrain may interfere between nodes or nodes and gateway communication. A wireless sensor node comprises four basic modules, namely sensor/actuator module, communication module, processing/computation module and power module (**Figure 6**).

A gateway is an interface between the application platform and the nodes on the WSN [17]. In a mesh network, any of nodes can be a gateway node as long as it has an external communication capability. A secondary, tertiary or even another primary gateway can be placed in the same network.

3.2. SeNoMa-Cloud Framework

The WSN management system enables various levels of management such as configuration, communication, performance and fault detection. In large-scale WSN deployments, it is crucial for the management system to enable self-healing and maintainability. The Sensor Node Management Cloud (SeNoMa-Cloud) proposed in [18] is a framework that centralises WSN sensor node management. SeNoMa was designed to address issues outlined for a WSN application named GeoSense [19] and evolved to support various types of WSN applications. The architecture complies with the Open Geospatial Consortium (OGC) Sensor Web Enablement (SWE) standard for sensor

http://dx.doi.org/10.5772/intechopen.70383

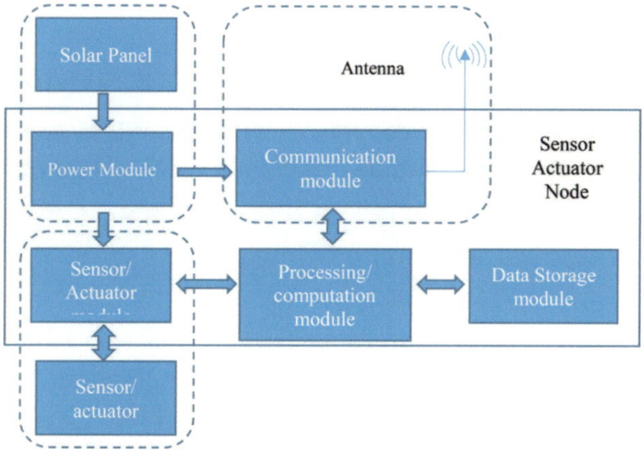

Figure 6. Basic components of a typical wireless sensor/actuator node.

data standardisation and supports communication protocols such as HTTP, MQTT and COAP. The three-layer SeNoMa-Cloud architecture is shown in **Figure 7** and is briefly described below.

3.2.1. Mote layer

This layer encapsulates the WSN and supports various gateway types, protocols and communication channels. For instance, a PC base station may utilise MQTT via an Ethernet connection to a remote server or an embedded resource-constrained gateway may use HTTP through GPRS.

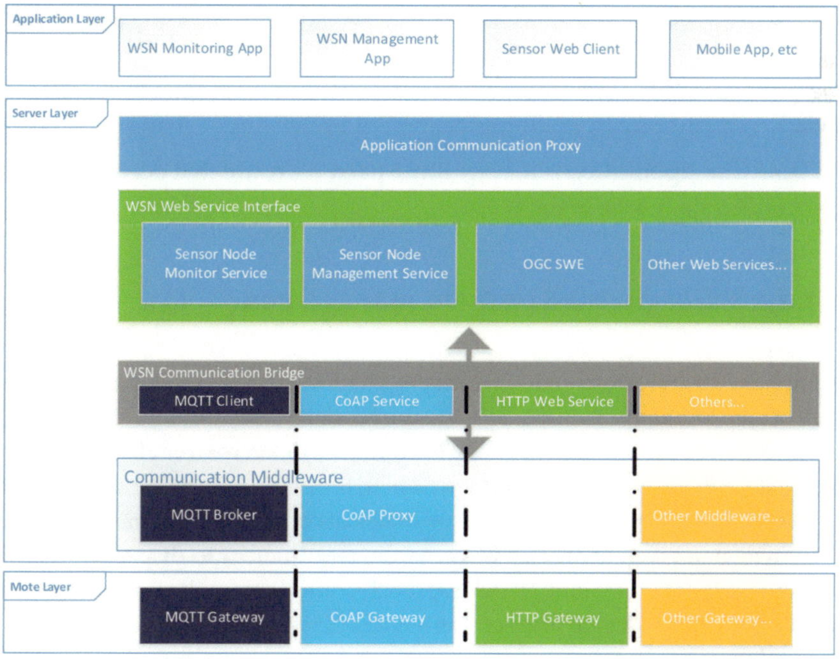

Figure 7. SeNoMa-Cloud Framework (adopted from [18]).

3.2.2. Server layer

This layer includes the services outlined below, including OGC SWE-compliant Web Services:

Sensor node monitor service: This service cooperates with the SWE to provide historical records of sensor data collected from data streamed live from a WSN communication bridge.

Sensor node management service: This service cooperates with the SWE to visualise the WSN sensor nodes and management interface, providing a mechanism to configure WSN node logging intervals.

WSN communication bridge: This accomplishes bidirectional communications between the WSN and Web services and is designed to support various communication protocols. It also handles encoding and decoding data communication from the WSN.

Communication middleware: This layer establishes communication with different types of WSN gateway, utilising different protocols and communication methods.

Application communication proxy: This relays requests utilising different protocols from an application layer to a corresponding service interface.

The server layer is provisioned for upgrades to include services such as notification or data modelling services.

3.2.3. Application layer

The application development depends on services provided by the server layer, such as the following applications:

WSN Monitoring App: This application is for monitoring and visualising sensor data from the sensor node monitor services. The monitoring application is designed based on client's requirements.

WSN Management App: This application allows network administrators to manage WSN functionalities via a web interface by communicating with sensor node management services.

Sensor Web Client: This application is for any third-party software that needs to access the SWE service to visualise data in geospatial maps.

4. Case study: WSN-enabled possum management

A small, light, easy to transport toxin delivery system has been developed to operate in the diverse range of environments that possums are found. Being small and light, many hundreds of devices can be deployed in problem regions. Each device delivers over 200 precisely measured doses of toxin to targeted animals and could remain in the environment for up to 2 years. Further successful field trials indicated the number of doses each device contained, far exceeded the number of animals in a given area. To overcome the requirement to dispense unused toxin till the next pest repopulation, a set and forget approach could remain the devices idle for months. Integrating a WSN into the device was seen as a cost-effective solution to monitoring and managing large number of devices.

With an ability to remotely monitor and gather data from WSN-enabled devices, conservation managers are able to employ more strategic approaches to animal identification,

http://dx.doi.org/10.5772/intechopen.70383

monitoring and trapping. Gathering device-specific data on the number of individual toxin deliveries, the remaining gas level for toxin shots and a time-stamp of triggering events enabled mapping of population densities and animal movements over time. In turn, this enables devices to be moved, removed or redeployed as required to more effectively manage possum. Although the opportunities to introduce additional sensors, for example environmental or atmospheric data gathering, the primary goal of this project was to enable close monitoring of the activity of each possum-specific toxin delivery device (actuator).

The case study presented herein is based on earlier work by Ghobakhlou et al. in [5] using Sensor Node Management Cloud (SeNoMa-Cloud) architecture for data monitoring and wireless sensor nodes management.

4.1. Prototype WSN system architecture

Wireless sensor actuator network (WSAN) is a WSN with an additional component, namely an actuator. This increases the capability of the WSN from simply monitoring to interactive control. A mesh network topology was used to develop a monitoring system for a possum-specific toxin delivery device (acting as an actuator). This network configuration enabled distribution of a number of devices over a large area, attaching sensor nodes to actuators and establishing internet communication via a GPRS-enabled master node. **Figure 8** illustrates mesh WSN configuration used for this deployment.

4.2. Field implementation

The prototype WSN was implemented and tested to carry data in a multi-hop network. The actuator module on the sensor nodes operated independently of the communication module. The events were logged and passed onto the storage module.

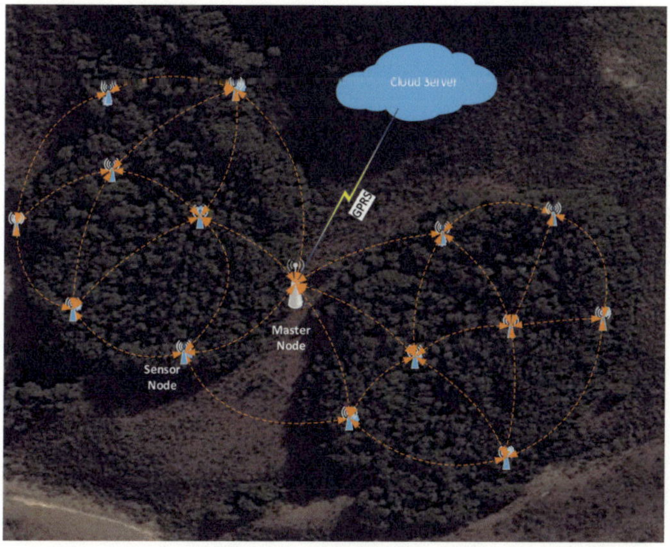

Figure 8. WSN mesh network architecture.

In this application, the WSN was intended to transmit log events to the master node 2 hourly and then returns back into sleep mode to preserve battery power. Nodes were set to transmit and receive logged data at 2-hourly intervals consuming relatively low power with included photovoltaic panels for continuous recharging. The proposed solution employed Waspmote [20] with ZigBee protocol, which uses the 802.15.4 standard and operates on the 2.4 GHz frequency, which falls under a licence-free frequency band in New Zealand.

The sensor/actuator module communicates with the sensor device, activates/deactivates the actuator and transmits data through communication device (e.g. ZigBee) to the master (sink) node. Sensor data from all nodes gathered in the master node and transmitted to a remote server via GPRS network according to predefined intervals. In this case study, the hardware specifications are illustrated in **Table 2**.

The acquired sensor data are processed by a microprocessor-based radio frequency (RF) device. The data, now in digital form, are packetized and dispatched to the central repository. **Figure 9** illustrates a proposed actuator/node prototype.

A trial series of 14 sensor nodes with WSN-enabled possum-specific toxin delivery devices was deployed and field-tested. The web interface (**Figure 10**) shows the results of these trials. Devices were monitored over a 6-week period. Two-hourly transmissions recorded the status for each node within the network. It shows data demonstrating time and frequency of toxin delivery events, toxin shots remaining and high battery power levels for all nodes despite having overcast weather for

Sensor node	Master node
8 MHz processing power	16 MHz processing power
2GB SD card	2GB SD card
1.8 W solar panel	1.8 W solar panel
XBee ZB Pro S2	XBee ZB Pro S2
6600 mAh battery	6600 mAh battery
	GPRS SIM928A

Table 2. Hardware node specifications.

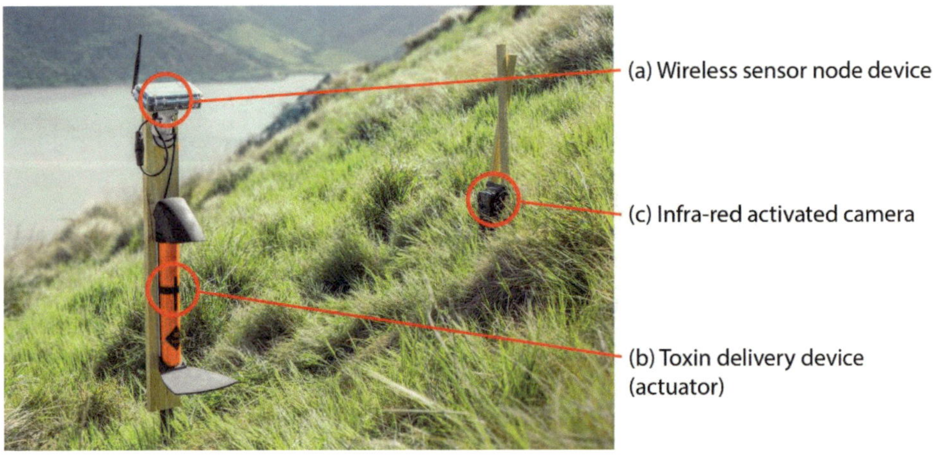

Figure 9. (a) Wireless sensor node device (b) An actuator/Toxin delivery device (c) Infra-red activated camera.

http://dx.doi.org/10.5772/intechopen.70383

the given period, demonstrating consistent packet delivery. This observation was based on power-saving mode where nodes went to sleep mode while waiting for the next transmission interval.

A web-enabled sensor node management's control panel was developed to allow remote configuration for each node. This application allows two-way communication between the server and WSN for multiple stations (see **Figure 11**).

Possum Control Project

Name	Last Record	Battery Power	Toxin Level	Gas Level	Total Fired	Last Fired Date
GATEWAY	9th Nov 2015, 15:00	91 %	N/A	N/A	N/A	N/A
REPEATER	26th Oct 2015, 03:00	93 %	N/A	N/A	N/A	N/A
NODE 1	22th Oct 2015, 03:00	97 %	48	118	3	9th Oct 2015, 11:36
NODE 2	28th Oct 2015, 17:00	100 %	48	119	13	11th Oct 2015, 04:56
NODE 3	23th Oct 2015, 07:00	98 %	49	119	7	11th Oct 2015, 02:38
NODE 4	25th Oct 2015, 23:00	98 %	46	116	3	12th Oct 2015, 21:56
NODE 5	22th Oct 2015, 01:00	98 %	68	138	0	Not Available
NODE 6	22th Oct 2015, 01:00	96 %	58	128	6	12th Oct 2015, 02:41
NODE 7	27th Oct 2015, 15:00	97 %	63	133	3	25th Oct 2015, 02:41
NODE 8	22th Oct 2015, 01:00	98 %	61	131	0	Not Available
NODE 9	25th Oct 2015, 13:00	98 %	63	133	1	20th Oct 2015, 04:07
NODE 10	27th Oct 2015, 21:00	98 %	63	133	1	11th Oct 2015, 21:51
NODE 11	22th Oct 2015, 01:00	94 %	62	132	0	Not Available
NODE 12	24th Oct 2015, 13:00	96 %	55	125	4	26th Aug 2015, 21:52

Figure 10. WSN's monitoring interface for possum control project.

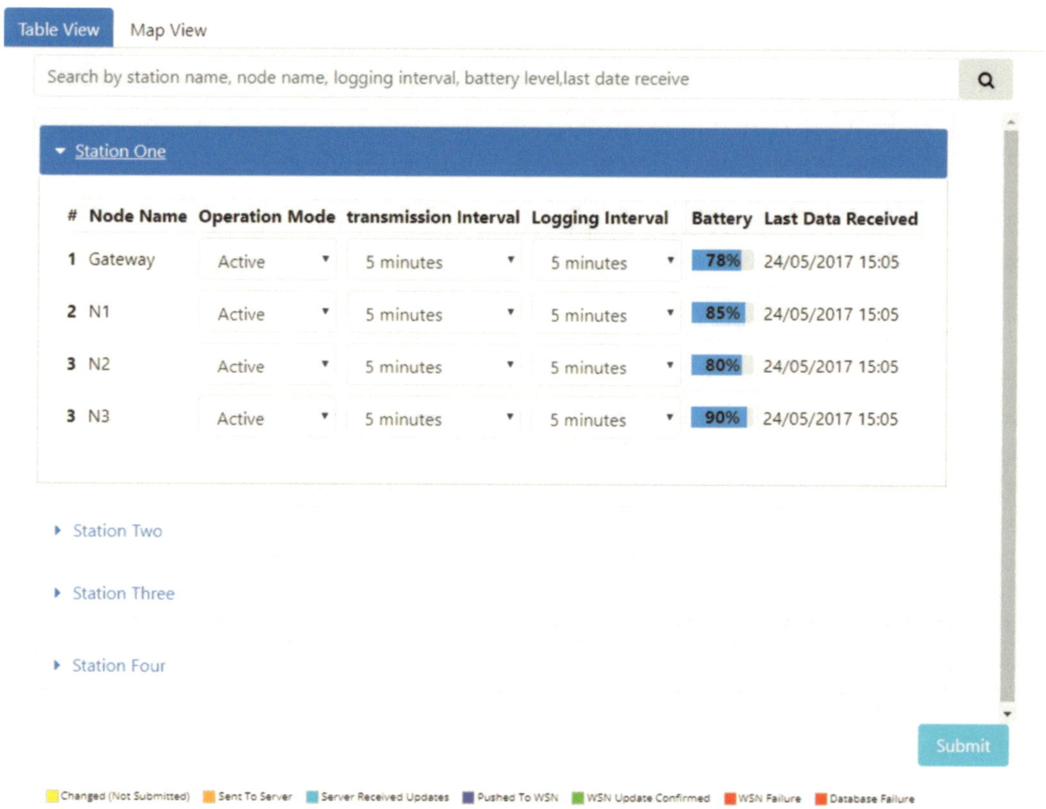

Figure 11. Sensor node management's control panel.

Figure 12. Map view of monitoring application interface.

The map view in **Figure 12** shows the spatial distribution of nodes along with number of times each node was triggered. A mouse over option displays a summary of statistics of the latest data such as node's remaining battery power, toxin and gas levels.

5. Conclusions

This chapter described opportunities for WSN in conservation management applications and presented WSN architecture for remotely monitoring pest control devices. The broad range of benefits for wireless sensor networks make them ideally suited to environmental monitoring and pest control. These include the ability to cost-effectively gather large volumes of long-term data and minimise labour hours in the field. In addition, an ability for nodes to save power by going into sleep mode when there is no activity and supplementing battery power by harvesting ambient solar energy gives them a long field life.

A WSN system monitoring a series of toxin delivery devices in a recent trial proved very successful and underlined the increased efficiency a reliable remote monitoring system could deliver. Conservation managers concluded that measures to decrease the necessity of manually checking devices in the field was a significant advantage and that the ability to remotely access the data reduced the labour and cost implications of the toxin delivery device trial. Further, an ability to turn off devices during periods of inactivity added a level of safety as well as increasing battery life. The sensor network enabled devices to be deployed, monitored and interactively managed, providing a significantly enhanced tool for the strategic management of pest species going forward.

Additional sensors in wireless sensor networks could also be used for monitoring other operational parameters at minimal extra cost. In addition to the primary device, data from a variety of sensors such as infrared to detect animal presence or humidity and barometric sensors for environmental monitoring could be added to the telemetry. The WSN could potentially decrease the operational costs of terrestrial monitoring and wildlife trapping programs significantly, particularly those involving labour-intensive manual checking.

Author details

Akbar Ghobakhlou* and Shane Inder

*Address all correspondence to: akbar.ghobakhlou@aut.ac.nz

Auckland University of Technology, Auckland, New Zealand

References

[1] Dowding JE, Murphy EC. The impact of predation by introduced mammals on endemic shorebirds in New Zealand: A conservation perspective. Biological Conservation. 2001;99(1):47-64

[2] Livingstone PG, Hancox N, Nugent G, de Lisle GW. Toward eradication: The effect of *Mycobacterium bovis* infection in wildlife on the evolution and future direction of bovine tuberculosis management in New Zealand. New Zealand Veterinary Journal 2015;63(supp 1):4-18

[3] Hutchings SA, Hancox N, Livingstone PG. A strategic approach to eradication of bovine TB from wildlife in New Zealand. Transboundary and Emerging Diseases. 2013;60(s1):85-91.

[4] Root-Bernstein M, Ladle RJ. Conservation by design. Conservation Biology 2010;24:1205-1211. DOI: 10.1111/j.1523-1739.2010.01501.x

[5] Ghobakhlou AX, Wang P. Sallis S. Inder and Blok S. Using WSN for possum management. In Sensing Technology (ICST), 2015 9th International Conference on, pp. 689-693. IEEE, Auckland, New Zealand, December 2015

[6] Jones C, Warburton B, Carver J, Carver D. Potential applications of wireless sensor networks for wildlife trapping and monitoring programs. Wildlife Society Bulletin 2015;39:341-348. DOI: 10.1002/wsb.543

[7] Gillies CA, Williams D. DOC Tracking Tunnel Guide v2.5.2: Using Tracking Tunnels to Monitor Rodents and Mustelids. Hamilton, New Zealand: Department of Conservation, Science & Capability Group; 2013

[8] Green W, Rohan M. Opposition to aerial 1080 poisoning for control of invasive mammals in New Zealand: Risk perceptions and agency responses. Journal of the Royal Society of New Zealand. 2012;42(3). DOI: 10.1080/03036758.2011.556130

[9] Department of Conservation. Battle for our Birds: Predator Control Programme [Internet]. 2014. Available from: http://www.doc.govt.nz/battleforourbirds [Accessed: May 2017]

[10] Department of Conservation. Department of Conservation Four Year Plan; Budget. July 2016. Available from: http://www.doc.govt.nz/Documents/about-doc/four-year-plan/doc-four-year-plan-2016.pdf

[11] Palma ART, Gurgel-Gonçalves R. Morphometric identification of small mammal footprints from ink tracking tunnels in the Brazilian Cerrado. Revista Brasileira de Zoologia. 2007;24(2):333-343

[12] Blackie H. www.landcareresearch.co.nz [Internet]. 2013. Available from: https://www.landcareresearch.co.nz/__data/assets/pdf_file/0004/66487/Blackie_Novel_automated_pest_detection.pdf [Accessed: 2017]

[13] Blackie H, MacKay J, Allen W, Smith D, Barrett B, Whyte B, Murphy E, Ross J, Shapiro L, Ogilvie S, Sam S, MacMorran D, Inder S, Eason CT. Innovative developments for long-term mammalian pest control. Pest Management Science. 2014;70(3):345-351

[14] Paek J, Hicks J, Coe S, Govindan R. Image-based environmental monitoring sensor application using an embedded wireless sensor network. Sensors. 2014;14(9):15981-16002

[15] Chen H-Y, Kuo Y-Y. Calculation of radio loss in forest environments by an empirical formula. Microwave and Optical Technology Letters 2001;31,no. 6: 474-480

[16] Ruhnau Pollak A. Challenges and Considerations for a Delay-Tolerant Wireless Sensor Network Deployment. 2010

[17] Cecílio, José, and Pedro Furtado. Wireless Sensors in Heterogeneous Networked Systems: Configuration and Operation Middleware. Springer, 2014

[18] Ghobakhlou, Akbar, Alexander Kmoch, and Philip Sallis. Integration of wireless sensor network and web services. In Proceedings of the 20th International Congress on Modelling and Simulation, Adelaide, Australia, vol. 16. 2013 (pp.838-844)

[19] Ghobakhlou A, Sallis P, Wang X. A service oriented wireless sensor node management system. In: Instrumentation and Measurement Technology Conference (I2MTC) Proceedings, 2014 IEEE International. Montevideo, Uruguay; May 2014. pp. 1475-1479

[20] Libelium, Waspmote Datasheet. www.libelium.com [Accessed: May 2017]

Chapter 6

Recent Advances on Implantable Wireless Sensor Networks

Hugo Dinis and Paulo M. Mendes

Additional information is available at the end of the chapter

http://dx.doi.org/10.5772/intechopen.70180

Abstract

Implantable electronic devices are undergoing a miniaturization age, becoming more efficient and yet more powerful as well. Biomedical sensors are used to monitor a multitude of physiological parameters, such as glucose levels, blood pressure and neural activity. A group of sensors working together in the human body is the main component of a body area network, which is a wireless sensor network applied to the human body. In this chapter, applications of wireless biomedical sensors are presented, along with state-of-the-art communication and powering mechanisms of these devices. Furthermore, recent integration methods that allow the sensors to become smaller and more suitable for implantation are summarized. For individual sensors to become a body area network (BAN), they must form a network and work together. Issues that must be addressed when developing these networks are detailed and, finally, mobility methods for implanted sensors are presented.

Keywords: implantable medical devices, sensors, communication, powering, mobility

1. Introduction

Implantable electronic devices are becoming ever smaller and more efficient, which drives their suitability for many new applications to levels never seen before. Examples of such devices are implantable chemical sensors [1], glucose and oxygen sensors for diabetics [2], neural implants [3, 4] and cochlear implants [5]. The constant evolution of these devices is paving the way for their large-scale use in the human body. It is not hard to imagine a cluster of sensors gathering data from several different locations in the human body, giving birth to what is referred to as a body area network (BAN). A BAN is a wireless sensor network (WSN) that consists of devices operating in, on or close to the human body [6]. It is composed of a small number of devices, equipped with biomedical sensors and wireless communications [7].

BANs and WSNs share many of the same challenges, but BANs pose a particular set of problems to be addressed:

- Size constraints imposed by the limited available space inside the body.
- Lossy materials surrounding the implant that heavily attenuate electromagnetic signals used for communication, degrading the quality of the link.
- Biocompatibility concerns.
- High power efficiency requirement, due to the limited available energy, whether from batteries or other powering methods [8].
- Communications must be reliable, as they can be conveying urgent information about life-threatening conditions of the individual.
- Data safety must be guaranteed during communication, as personal and confidential medical information will be transmitted.

In this chapter, implantable sensors and sensor networks will be studied, starting with examples of their applications in the biomedical field and state-of-the-art sensors. Methods that allow the devices to communicate with the outside world will be reviewed and discussed, as the sensors must transmit data to an external reader, so that it can be accessed by the individual or medical personnel. Alternative powering methods that allow the device to have smaller form factors than those possible with batteries will be presented. Advances in material science and fabrication techniques lead to the integration of electronics with smart materials, thus birthing a new generation of devices that are more suitable than ever for implantation. These integration efforts will be presented in this chapter. As the main interest of this publication is wireless sensor networks, networking issues faced by in-body sensors will be presented. Finally, there have been reports of self-propelled devices, and the possibility of having sensors capable of moving to different places in the human body inspired the authors to present some of these propulsion methods.

2. Applications

The biomedical field has a vast range of devices and techniques capable of aiding medical staff to diagnose, manage and treat diseases. This section focuses on sensors, which are responsible for gathering data on a given biomedical signal and relaying them to physicians. Since the scope of this book is wireless sensor networks, only sensors with wireless capabilities will be considered and presented. Examples of applications of sensors in the medical field will be presented, along with proposed devices.

2.1. Intraocular pressure

Intraocular pressure (IOP) monitoring is an important tool for medical staff to diagnose and control glaucoma. This disease is the second most common cause of blindness, and it is predicted to affect around 80 million people by 2020 [9]. Different approaches for measuring IOP are possible and range from non-invasive devices, such as contact lenses [10, 11] that measure the deformation of the cornea curvature due to the extra pressure, to invasive, implantable sensors [12–16] that directly measure the IOP inside the eye. The device presented in Ref. [12] is considered by the authors of this chapter to be the state of the art of IOPs. It is a 1.5 mm^2 sensor with wireless communication capabilities and a power requirement of only 7 μW, which is satisfied by solar energy harvesting.

2.2. Neural activity

Neural activity provides useful data for a number of different applications. It can be used, for example, to diagnose neural dysfunctions, such as epilepsy [17], to control prosthetic limbs through what is called a brain-machine interface [18], and behavioural studies [19]. Neuron action potentials can be measured from deep brain tissues through implantable needles [20], or from the surface of the cortex [17, 21–23], which reduces cortical scarring and allows for chronical and stable measurements [21]. There are even examples of devices used to record the electrical activity of neuron in the peripheral nervous system [18]. The state of the art in implantable neural sensors is considered to be the device presented in Ref. [21]. It is a radio frequency (RF) wirelessly powered, 42.25 mm^2, 64-channel sensor with a 1 Mbps data rate, consuming only 225 μW. A smaller device consuming 120 μW is available in Ref. [18], but its single-channel topology puts it at a disadvantage.

2.3. Bladder pressure

Bladder pressure monitoring is an important tool for the diagnosis of bladder dysfunctions. As some symptoms may only be induced in normal daily activities, such as walking, they cannot be registered in an acute measurement at the hospital. Implantable, chronic reading is necessary, preferably with no discomfort to the patient. Examples of such devices are presented in Ref. [24–27]. In Ref. [27], a 40 mm^2 sensor consuming 16 μW, with sound wave power transfer capabilities and LC resonance-based communication, is presented and considered to be the state of the art in this field.

2.4. Glucose

Glucose monitoring is traditionally done by the patient himself, usually by pricking the fingertip and drawing a small blood sample. Unfortunately, this method is not comfortable for the patient and is only capable of getting a measurement in given points of time. Implantable alternatives are being researched and have already been presented [28, 29]. These allow for continuous glucose-level monitoring and can be used to trigger alarms or even to automatically control implantable insulin pumps, thus improving the patient's quality of life. In Ref. [28], a needle implantable $0.5 \times 0.5 \times 5$ mm^3 wireless sensor with light powering and communication is presented.

2.5. Blood pressure

High blood pressure is the main cause for morbidity and mortality worldwide [30]. It is responsible for a higher risk of cardiovascular diseases, heart problems, strokes and aneurisms. Being such a critical vital parameter, continuous monitoring can prove important to the medical staff when it comes to diagnosing conditions. Implantable wireless blood pressure sensors have been proposed in Ref. [30–33]. Blood pressure can also be useful to control vascular graft degradation through blood flow measurements, and a sensor capable of performing this task is presented in Ref. [34]. The state of the art is considered to be the vascular graft blood pressure sensor presented in Ref. [34]. This sensor has a 2.67 mm^2 chip with two coils that hold it in place inside a vascular graft. Pressure is digitized and backscattered, with the device consuming only 21.6 μW and with a sensitivity of 0.176 mmHg.

2.6. pH

The pH of a solution plays an important role in chemical processes that it undergoes, therefore affecting several physiological parameters and functions. pH can be used to identify microbial presence in tumours and monitor wound healing [35]. In Refs. [36, 37], a sensor is used to monitor gastroesophageal reflux disease (GERD) by measuring pH in the oesophagus. In Ref. [38], oral pH is measured to control the pathogenesis of dental caries. The device presented in Ref. [35] is the state of the art of implantable pH sensors. It integrates carbon nanotube-based sensors, which do not require a reference electrode, with an RFID tag that modulates data into an externally provided carrier. It is capable of accurately detecting pH levels between 2 and 12 during 120 days.

2.7. Intracranial pressure

Intracranial pressure is a vital biomedical parameter when it comes to the management of traumatic brain injuries. Current methods require catheters inserted into the cranial cavity, which cause patient discomfort and carry a risk of infection and haemorrhage [19, 39]. Minimally, invasive techniques based on wireless sensors have been presented in Refs. [39–42]. Chen et al. presented, in Ref. [39], passive sensors with volumes down to $1 \times 1 \times 0.5$ mm^3. Pressure ranges from 0 to 100 mmHg were registered, with wireless and batteryless operation.

2.8. Electromyography

Electromyography (EMG) measures the electrical potentials present in muscle, and this data can be useful for the diagnosis of illnesses and injuries, functional electrical stimulation, and to control prosthetic limbs. EMG sensors with wireless capabilities have been presented in Refs. [43, 44]. The sensor presented in Ref. [44] is an EMG and electrocardiogram (ECG) monitor with four analog channels, a chip that consumes 19 μW (when sampling from one channel) and communicates at a data rate of 200 kbps with a power consumption of 160 μW. It includes RF power transfer and thermoelectric energy harvesting powering modules, giving the device versatility.

2.9. Electrocardiogram

Electrocardiogram (ECG) measurements allow physicians to have a closer look at the patients' heart, and it can be used to detect arrhythmias and heart attacks (myocardial infarctions), for example. Wireless ECG monitors have been proposed in Refs. [44, 45]. The device presented in Ref. [45] is notable for its extremely low power consumption of 64 nW, which raises the bar in terms of power budgets. Nevertheless, it does not allow for continuous monitoring, as it only stores abnormal events into the memory for posterior wireless relaying. For continuous monitoring, the previously discussed device of Ref. [44] is considered the state of the art.

3. Communications

Promising and viable communication strategies have been reported, such as intra-body communication (IBC) [46, 47] and ultrasound (US) [48]. The first consists on using biological tissue of the system's host as a conductive medium for electrical signals conveying data. The second is based on ultrasounds, a mechanical wave of frequencies above 20 kHz, which suffers low tissue absorption. Radio frequency (RF) is the most widely implemented communication method; therefore, this section will focus on sensors with RF wireless communications. Passive and active RF communication methods will be presented, with examples of devices resorting to them.

3.1. Passive RF communication (PRFC)

This communication method relies on the resonant frequency of a pair of coupled coils, one in the wireless implant and the other in an external device. The sensor is attached to the implant's coil, and a change in the parameter to which the sensor is sensitive to translates into a varying impedance of the coil. Consequently, the resonant frequency of the coupled coils will shift as the parameter of interest, for example, IOP, varies. Generally, this approach requires no power from the implant [13, 14, 32, 33, 39, 40], as the external reader is responsible to detect the impedance change in the implant's coil and, from it, calculate the sensed parameter's value. This allows for smaller implants, as power budget is reduced and no processing electronics are required. In a BAN perspective, this communication method can be applied in situations where on-body readers are a possibility (e.g. intraocular pressure monitoring where the external reader is placed in a pair of glasses worn by the patient).

3.2. Active RF communication (ARFC)

Implantable sensors described in this subsection communicate with the outside world resorting to an on-board antenna and an RF signal, at the expense of power. In Ref. [30], the authors resorted to capacitive coupling, in contrast to the more common inductive coupling. This method consists of using the host's biological tissue as a dielectric between two sets of metallic plates, one on the sensor and another on a reader, which can be body-worn or implantable. Operation frequencies must be kept as low as possible, since the tissue becomes more conductive as frequency increases.

Inductive coupling communication is performed between two coils and has the advantage of being more efficient than far-field communication. On the other hand, a precise alignment between coils is necessary, under the penalty of drastically losing efficiency. Additionally, the distance between coils must be kept as small as possible, unlike far field, which can be used at long ranges. The choice between one of these two powering methods must be made considering the available space, power budget and radiation safety guidelines. From the examples provided, no connection can be made between the choice of inductive coupling or far-field communication and the application of the sensor.

The choice between passive and active communication technologies is one that cannot be taken lightly when designing and developing an implantable sensor. Considering BANs, if wearable or large implantable relays are available near the implanted sensors, and the latter have severe volume limitations, passive communication can be a viable option, as the relays can support the bulky batteries or wireless power transfer (WPT) components, and the implanted sensor can use an oscillator to modulate the data into the relay's RF signal and backscatter it. In applications where the sensor has available space for computing capabilities, inductive links can be employed. With this, the sensor can process larger amounts of data, such as multiple channels. When long-distance operation is desirable, e.g. when no on-body or implantable relays are desirable, far-field communication is the best option, as it removes those constraints.

4. Powering

Sensor miniaturization is a desired goal; therefore, a compromise must be made between battery size, and consequently the size of the device itself, and its autonomy, bearing in mind that battery replacement may require an invasive surgical procedure, which could potentially lead to health complications [49, 50]. The urge to research for new and reliable powering solutions for implantable devices to increase their lifespan and reduce their volume is evident, and the interest in this field is proven by the amount of publications made available over the previous years. **Figure 1** contains a diagram representation of the different types of device powering that will be discussed in this section.

4.1. Energy harvesting

Energy harvesting techniques consist of harvesting useful amounts of energy from the ambient environment in order to power a device or charge a battery, having potential to provide power to biomedical devices since they could yield unlimited energy, drastically increasing the devices' lifespan. However, harvesting useful amounts of energy from the environment can be proven challenging, as the amount of available energy is volatile and often very limited, which imposes the need of special power management circuitry [49, 51]. Despite of the aforementioned limitations, research in the field of energy harvesting is of high interest due to the constant reduction of the power demands of electronic circuits [52].

Several energy harvesting techniques have been proposed by researchers, and special attention is given to thermoelectric generators, biomechanical energy, solar power, biofuel and RF energy harvesters.

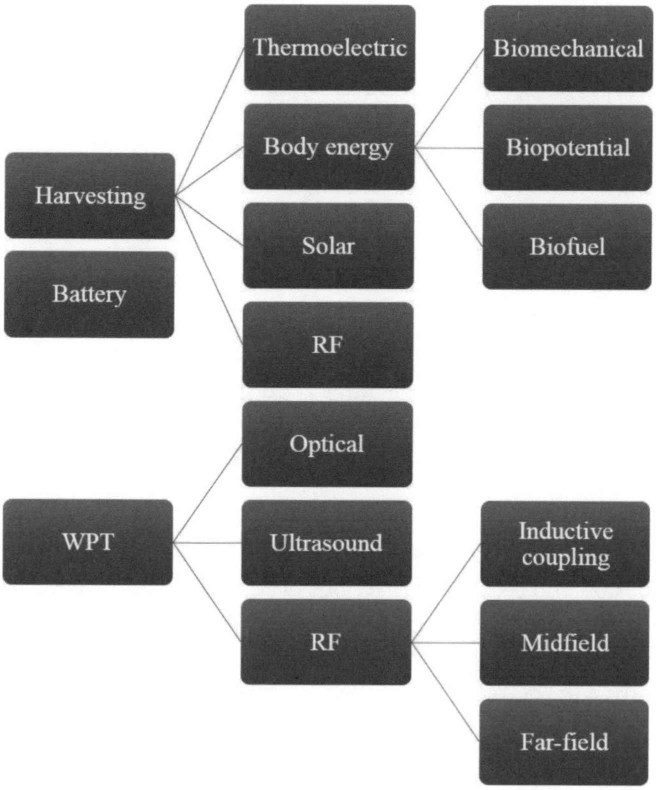

Figure 1. Implantable device powering methods.

4.1.1. *Thermoelectric generators*

Thermoelectric generators are solid-state devices that convert the thermal energy from temperature gradients into electrical energy [53]. These are appealing power sources for implantable devices, as they possess high reliability, are compact and do not require moving parts [54]. Such generators are based on the Seebeck effect, which states that an electrical voltage is generated across a metal or semiconductor when it is exposed to a temperature gradient [55].

4.1.2. *Biomechanical energy harvesters*

Biomechanical energy is generally abundant in the human body. It is generated by breathing, muscle stretching, body weight during motion and heart beats. The conversion between energy types is achieved resorting to a transduction mechanism, with electromagnetic and piezoelectric mechanisms being the most common. Biomechanical energy harvesters usually fall into two categories: vibrational or force-driven. Vibrational harvesters use inertial energy of a given mass, while force-driven ones rely on direct application of mechanical force [49].

This generator's feasibility for implantable medical devices was also studied. It was implanted on the right ventricular wall of a dog's heart and produced 80 mJ of energy after 30 minutes of operation [56].

4.1.3. Solar power

Solar cells were found to be capable of powering implantable devices. Even when implanted below a skin layer, these cells can harvest some power, as a small amount of light is able to penetrate the skin, in particular near-infrared light [57]. An absorption of around 10% of the incident power per millimetre of the skin occurs for a wavelength of 632.8 nm and 11.5% for 904 nm [58]. Nevertheless, large size, low efficiency and tissue heating are the major drawbacks of these systems [59]. Solar power harvesting cells have been developed in Refs. [12, 60], and they are capable of generating 1.1 $\mu W/mm^2$ in the eye and 34 $\mu W/mm^2$ below the skin, respectively.

4.1.4. Biofuel

Biofuel cells transform biochemical energy into electric energy by making use of electrochemical reactions. Oxidation and reduction reactions occur in the anode and cathode of the biofuel cell, generating a flow of electrons that generates power that a device can be used to power itself. Advantage of this technology is, for example, the biocompatibility between the fuel cell and the human body. On the other hand, low harvested power levels can pose a limitation as well as the anode and cathode degradation over time. Examples of reviews of implantable biofuel cells in living animals are available in Refs. [61, 62].

4.1.5. RF energy harvesting

RF energy harvesting consists on harnessing electromagnetic waves that exist in the environment, generated by communication towers, for example. These waves have the potential to provide power for electronic devices. The quantity of available radiation, the efficiency of the power conversion system and the size constraints of the device will dictate whether this method suffices in powering a given application. Even though technological advances are constantly being made, the size constraint of implantable medical devices and the typical ambient RF power densities cause some uncertainty about the suitability of this device powering method, as power levels below 1 μW can be recovered [51].

4.2. Wireless power transfer

The previously studied energy harvesting techniques suitable for implantation generate small power outputs. Consequently, the use of a dedicated power emitter for charging the devices has to be considered. Pertinent technologies such as the use of optical energy, ultrasounds or RF waves emerge as alternatives.

4.2.1. Optical link

Optical waves have been suggested to power medical implants, as they do not interfere with nearby communication systems like RF waves do. In Ref. [63], an array of silicon diodes with an area of 2.1 cm^2, implanted 1–3 mm under the skin, was used for transcutaneous power transmission. Using near-infrared irradiation at 810 nm with a power density of 22 mW/cm^2, the charging of a lithium battery capable of powering a commercial pacemaker for 24 hours was reported, while the temperature rise on the skin during light irradiation was 1.4°C.

4.2.2. *Ultrasonic link*

Ultrasonic waves, akin to optical waves, do not interfere with nearby electromagnetic fields and communication devices. They induce a vibration in the tissue, and the resulting kinetic energy is converted to electrical energy through a transducer, e.g. a piezoelectric transducer [49]. Ultrasonic power transmission has some disadvantages that limit its application to implantable medical devices. This transmission is very sensitive to the contact between the transmitter and the tissue, as an impedance mismatch between these elements or a misalignment between transmitter and receiver can severely reduce transmission efficiency [64].

4.2.3. *Radio frequency link*

Electromagnetic radiation, more specifically RF, is adequate to transport energy over long distances and presents one of the highest miniaturization potentials [65]. Additionally, its absorption by biological tissues does not induce damage, as long as the specific absorption rate (SAR) is not exceeded.

One of the most common methods of power transmission to medical devices bases itself on inductive coupling, as it has the lowest absorption rate by body tissue at lower frequencies. This method has been previously used to power cochlear implants, total artificial hearts and neural implants, among others [66–69]. Despite its popularity, this method has some drawbacks, such as coil decoupling due to misalignment, since it requires rigorous positioning of transmitter and receiver coils [70–72]. Moreover, the range of inductive coupling complies with exponential decay, a near-field phenomenon, meaning that the external coil must be close to the implant. These limitations can be overcome by establishing links in the middle (see Ref. [73]) or far field, resorting to antennas. Although energy transportation is less efficient, it allows for greater distances between the power source and the target than the previous inductive methods [74].

5. Integration

Most of today's implantable electronic devices, such as the ones so far reviewed in this chapter, rely on silicon microelectronics. The evolution of fabrication techniques and microelectronics has translated into a reduced size of implantable electronics. Nowadays, there is an urge to further miniaturize them to make them easier to implant and less traumatic for the patient. Efforts made towards this goal over the past few years will be reported through examples of success cases.

Chen et al. [39] have designed a wireless pressure monitoring sensor with dimensions down to $1 \times 1 \times 0.1$ mm^3. A $2.5 \times 2.5 \times 0.1$ mm^3 device was used to validate the design in vivo. A resonant circuit composed of an inductive antenna and a pressure-sensitive capacitor is the heart of the sensor. An applied pressure changes the resonant frequency of the LC circuit, and the frequency shift is detected by an external reader, which then converts it into pressure values.

Mostafalu et al. [75] created threads with different properties to act as sensors, microfluidics and electronics. Hydrophobic threads were used as microfluidic channels, while threads infused with materials such as carbon nanotubes were used as electrodes for sensing pH, glucose and so on. Conventional electronics were present in a different layer, and these established communication links and processed the electrodes' signal. Fabric devices were tested in pH and strain sensing, in vivo, having been successful. This research has the potential to lead to the creation of smart sutures and bandages.

The examples above serve to demonstrate how fabrication technology enables devices to become smaller than ever, while still packing enough features to perform their given tasks.

5.1. Biodegradable and stretchable sensors

In the past few years, new materials for implantable sensors have been proposed, studied and validated, namely, stretchable and biodegradable materials [76–80]. Biodegradable materials allow for transient sensors that can, for example, be implanted after a surgery to monitor wound healing and bacterial activity, and after a predefined period, the device would start to degrade inside the human body. The by-products of this process would then be eliminated naturally by the organism. This would mitigate the need for implant retrieval surgeries, along with all associated negative aspects, e.g. patient discomfort, risk of infection, surgery room scheduling and so on.

Kang et al. [41] demonstrated an ICP sensor in a rat, fabricated with a polymer (PLGA) and either a magnesium or a silicon foil. Continuous monitoring of ICP was achieved during 3 days, after which the materials composing the sensor were reabsorbed into the body.

Luo et al. [81] fabricated a pressure sensor based on a variable capacitor and a coil. The biodegradation of this device was documented by the authors. During the first 21 hours of immersion in a saline solution, the resonant frequency of the sensor changed, as if it was stabilizing itself in the system. In the following 86 hours, the resonant frequency stayed constant, showing stability of the device, thus being the optimal operation period of the sensor. After this, the quality of the sensor starts to degrade until it is unusable.

A transient device capable of managing bacteria growth in a region of the body, possibly a surgery or implant site, has been proposed in Ref. [80]. Using magnesium for an inductive coil, a silicon resistor and silk encapsulation, a heater was produced. An external RF field would power the resistor, which would heat up by 5°C and prevent bacteria proliferation in that location. The longevity of the device is controlled by the silk's crystallinity.

Biodegradable batteries have also been achieved (see Ref. [78]). These were capable of powering a LED and a 58 MHz wireless signal generator.

Finally, stretchable electronics are also becoming a reality [79]. A sensor that can be bent and twisted without losing its properties is an important step in implantable devices, as patient discomfort would be greatly reduced. Devices, such as electronic eyeball cameras and coplanar waveguides, have been demonstrated.

The evolution of integration techniques and material science is of paramount importance for the medical sensor area. Smaller devices with the same powerful capabilities are in high demand, allowing for new applications and the improvement of current ones. Bendable and stretchable sensors can be a positive step in patient comfort and device reliability, reducing the negative response from the human body. Finally, biodegradable sensors have an enormous potential, as they can be used to monitor a parameter for a limited period of time, after which it is simply absorbed by the human body without any harm, eliminating the need for a retrieval surgery.

6. Networking issues

The increase of implantable sensor solutions for the medical field brings the necessity of such sensors to work together to collect and relay measurements of biomedical parameters. The most used communication method for implants is based on electromagnetic radiation. Due to the conductive nature of biological tissue, it suffers great attenuation, as tissues absorb energy and dissipate it as heat. Experimental path loss models were presented in Ref. [82] for in-body to in-body, in-body to on-body and in-body to off-body communications between 2.36 and 2.5 GHz. This work is a proof of the challenges that lossy biological tissue presents to sensor development and networking.

Networking solutions of implanted sensors must consider SAR limits and temperature increase of tissues to guarantee patient safety. According to Ref. [83], the high quality of service (QoS) required for biomedical systems can only be achieved in such a propagation medium if performance-enhancing techniques, such as adaptive coding and modulation and link diversity, are adapted from miniature wireless electronics to implantable sensors.

In Ref. [7], three networking methods for on- and in-body sensors are presented. Of these, the use of on-body beacons shows good promise. The beacons would be responsible for forwarding data between sensors and relaying it to base stations, thus reducing the power dissipation inside the human body. Since the beacons can be larger than implantable sensors, these can also be used as power sources or controllers for the sensors, as their power budget can be significantly higher. In Ref. [84], the authors agree with the previous statement, and they present a study of QoS and power consumption variation of BAN nodes with different on-body beacon placements.

BANs pose yet another challenge for engineers. As the human body is a flexible, moving environment, the relative position of the network's nodes can change frequently, thus altering signal attenuation in communication links. For example, a wearable, on-body relaying node, such as a smartwatch, changes its position relatively to the in-body sensors all the time during normal day-to-day activity of the wearer. The same concept applies to in-body sensors placed in moving organs and members or even in the bloodstream. Ramachandran et al. [85] proposed a medium access control (MAC) protocol based on human activity, which they named HAMAC, that aims to work around the previously mentioned problems by adjusting the timing of communication between nodes and from nodes to relays.

In 2012, the IEEE has published standard IEEE 802.15.6. It addresses the communication protocol for BANs for medical applications (see Ref. [6]). Works around this standard have been found, such as the one found in Ref. [86], which aims to improve it by adding an ultrawideband channel model.

Alternatives to RF networking have also been studied. Santagati et al. [48, 87] proposed a MAC protocol for US communications, ultrasonic wideband (UsWB). It aims to establish intra-body communication between BAN nodes without the previously mentioned setbacks of RF radiation. UsWB was reported to be resistant to the multipath caused by the multitude and inhomogeneity of tissues in the propagation medium, i.e. the human body, thus making it a viable alternative to RF-based communications.

6.1. Security concerns

When sensors transmit data to one another or to the outside world, sensitive medical information is therein contained. The theft of such data by a third party is a serious danger and must be prevented. In the case of sensors or actuators within the body network that receive instructions from a controller, the possibility of having an attacker sends commands to these devices must be completely eliminated to guarantee the safety of the patient. Furthermore, an attacker must not be able to modify the content of the data being exchanged in the network without the receiver noticing the change, thus guaranteeing the integrity of the communication.

Steps towards the protection and encryption of transmitted data have been taken and reported. In Ref. [88], the authors proposed a method to share secret data inside a network by using ECG as a decryption key. Only an external reader with access to real-time ECG of the patient would be able to read the data, and given the random nature of the ECG wave, this safety method presents great potential. Nevertheless, it must not be forgotten that implantable sensors have limited power budgets; therefore, this encryption must be lightweight. In Ref. [89], the same authors have improved upon this method by using characteristic parameters of ECG signals, the P, Q, R, S and T peaks, and generate random binary sequences with the time intervals between these peaks. This approach was reported to have low latency and to benefit of the same randomness of ECG signals as the previously reported one.

7. Node mobility

In recent years, propulsion methods for small implantable devices, or robots, have been proposed. Having sensors capable of moving in body fluids has medical interest, as it can allow one device to perform measurements and diagnostic in an area wider than ever before.

Minimally invasive surgery or targeted drug delivery could also be performed with steerable devices.

In Ref. [90], a 3 × 4 mm^2 wireless implantable device is presented. Its propulsion method is based on magnetohydrodynamics (MHD). It requires a constant magnetic field of around 0.1 T, which can be achieved with a permanent magnet. The device applies currents in the mA order of magnitude through the medium's conductive fluid, and a force is generated in the magnetic field. The device then experiences an equal and opposite force that propels it. Power is provided by a 1.86 GHz WPT system, which also carries movement commands modulated into the power carrier. The device can controllably move at a speed of 5.3 mm/sec in salt water.

Hsieh et al. [91] developed a remote-controlled device with a propulsion mechanism based on gas pressure from electrolytic bubbles generated on the surrounding fluid. It can move at a rate of 0.3 mm/s, at around 200 μW power consumption. Electrolysis electrodes are present all around the device, so it is possible to define where the electrolysis will occur and, consequently, steer the device. It is powered by a 10 MHz inductive coupling link which also carries commands to control movement direction and speed. The receiving coil and electrodes are integrated in the locomotive chip, which has a total area of 21.2 mm^2. Despite the slower speed of this device, especially comparatively to the one reported in Ref. [90], this approach does not require external components such as permanent magnets.

This section presented propulsion mechanisms for wireless devices operating in a liquid medium. The reported WPT, communication and steering capabilities of these devices are an important stepping stone towards fully autonomous or remote-controlled sensors and actuators integrating a BAN that are capable of navigating, for example, through the bloodstream, digestive tract or bladder. Simultaneously, they would be performing measurements, relaying them to the outside world and performing microsurgery or drug delivery at the required locations.

8. Conclusion

BANs comprised of implantable sensors are becoming closer and closer to being a common tool in the medical field. This would mean a significant improvement on healthcare for patients, as close monitoring of critical parameters can be done full-time and without constraints. Several powering methods that allow these devices to be as small as possible and to operate indefinitely are available and maturing. The same applies for communication methods, which tend to be less power consuming, and there were even reported completely passive methods that can be used in situations where extremely small devices are required. Still in the topic of communications, security issues and networking difficulties have been raised, with efforts to mitigate them being presented. Integration techniques that allow the fabrication of sensors with more host-friendly materials have been detailed, with biodegradable and stretchable materials being a topic of high interest in the past few years. Finally, mobility mechanisms that allow for controllable exploratory sensors have also been shown, and these pave the way for large area monitoring by a single sensor, adding to their versatility and capabilities. In conclusion, the evolution of biomedical sensors is leading the way to completely functional and tailored BANs that in the near future will prove to be indispensable tools for health monitoring in both the hospital environment and daily life of patients.

Acknowledgements

This work is supported by FCT with the reference project PTDC/EEI-TEL/5250/2014, by FEDER funds through Projecto 3599 – Promover a Produção Científica e Desenvolvimento Tecnológico e a Constituição de Redes Temáticas (3599-PPCDT) and by grant SFRH/BD/116554/2016.

Author details

Hugo Dinis* and Paulo M. Mendes

*Address all correspondence to: hugodcdinis@gmail.com

Department of Industrial Electronics, University of Minho, Portugal

References

[1] Frost MC, Meyerhoff ME. Implantable chemical sensors for real-time clinical monitoring: Progress and challenges. Current Opinion in Chemical Biology. 2002;**6**(5):633-641

[2] McKean BD, Gough DA. A telemetry-instrumentation system for chronically implanted glucose and oxygen sensors. IEEE Transactions on Biomedical Engineering. 1988; **35**(7):526-532

[3] Wise KD, et al. Wireless implantable microsystems: High-density electronic interfaces to the nervous system. Proceedings of the IEEE. 2004;**92**(1):76-97

[4] Olsson R, Wise K. A three-dimensional neural recording microsystem with implantable data compression circuitry. IEEE International Solid-State Circuits Conference. 2005;558-559. DOI: 10.1109/ISSCC.2005.1494117

[5] Bazaka K, Jacob M. Implantable devices: Issues and challenges. Electronics 2012;**2**:1-34 p

[6] IEEE Standards Association. IEEE Standard for Local and Metropolitan Area Networks—Part 15.6: Wireless Body Area Networks. IEEE Std. 2012. 271 p. DOI: 10.1109/IEEESTD.2012.6161600

[7] Honeine P, et al. Wireless sensor networks in biomedical: Body area networks. Systems, Signal Processing and their Applications (WOSSPA), International Workshop. 2011; (1):388-391. DOI: 10.1109/WOSSPA.2011.5931518

[8] Hannan MA, et al. Energy harvesting for the implantable biomedical devices: Issues and challenges. Biomedical Engineering Online. 2014;**13**(1):79

[9] Quigley HA. Number of people with glaucoma worldwide. The British Journal of Ophthalmology. 1996;**80**(5):389-393

[10] Chiou JC, et al. Toward a wirelessly powered on-lens intraocular pressure monitoring system. IEEE Journal of Biomedical and Health Informatics. 2016;**20**(5):1216-1224

[11] Leonardi M, et al. Wireless contact lens sensor for intraocular pressure monitoring: Assessment on enucleated pig eyes. Acta Ophthalmologica. 2009;**87**(4):433-437

[12] Ghaed MH, et al. Circuits for a cubic-millimeter energy-autonomous wireless intraocular pressure monitor. IEEE Transactions on Circuits and Systems I Regular Papers. 2013;**60**(12):3152-3162

[13] Chitnis G, et al. A minimally invasive implantable wireless pressure sensor for continuous IOP monitoring. IEEE Transactions on Biomedical Engineering. 2013;**60**(1):250-256

[14] Kouhani MHM, Weber A, Li W. Wireless intraocular pressure sensor using stretchable variable inductor. Proceedings on IEEE International Conference on Micro Electro Mechanical Systems. 2017. pp. 557-560. DOI: 10.1109/MEMSYS.2017.7863467

[15] Donida A, et al. A 0.036 mbar circadian and cardiac intraocular pressure sensor for smart implantable lens. Dig Tech Pap – IEEE International Solid-State Circuits Conference. 2015;**58**:392-393

[16] Shih Y, Shen T, Otis B. A 2.3 μW wireless intraocular pressure/temperature monitor. IEEE Journal of Solid-State Circuits. 2011;**46**(11):2592-2601

[17] Chen W, et al. A fully integrated 8-channel closed-loop epileptic seizure control. IEEE Journal of Solid-State Circuits. 2014;**49**(1):232-247

[18] Seo D, et al. Wireless recording in the peripheral nervous system with ultrasonic neural dust. Neuron. 2016;**91**(3):529-539

[19] Fan D, et al. A wireless multi-channel recording system for freely behaving mice and rats. PLoS One. 2011;**6**(7):1-9. DOI: 10.1371/journal.pone.0022033

[20] Rhew HG, et al. A fully self-contained logarithmic closed-loop deep brain stimulation SoC with wireless telemetry and wireless power management. IEEE Journal of Solid-State Circuits. 2014;**49**(10):2213-2227

[21] Muller R, et al. A minimally invasive 64-channel wireless μECoG implant. IEEE Journal of Solid-State Circuits. 2015;**50**(1):344-359

[22] Mestais CS, et al. WIMAGINE: Wireless 64-channel ECoG recording implant for long term clinical applications. IEEE Transactions on Neural Systems and Rehabilitation Engineering. 2015;**23**(1):10-21

[23] Gao H, et al. HermesE: A 96-channel full data rate direct neural interface in 0.13 μm CMOS. IEEE Journal of Solid-State Circuits. 2012;**47**(4):1043-1055

[24] Majerus SJA, et al. Wireless, ultra-low-power implantable sensor for chronic bladder pressure monitoring. ACM Journal on Emerging Technologies in Computing Systems. 2012;**8**(2):1-13

[25] Majerus S, et al. Wireless implantable pressure monitor for conditional bladder neuromodulation. IEEE Biomedical Circuits and Systems Conference: Engineering for Healthy Minds and Able Bodies, BioCAS 2015 – Proceedings. 2015. pp. 2-5. DOI: 10.1109/BioCAS.2015.7348337

[26] Lee WS, et al. UP-link: An ultra-low power implantable wireless system for long-term ambulatory urodynamics. 2014 IEEE Biomedical Circuits and Systems Conference (BioCAS) – Proceedings. 2014. pp. 384-387. DOI: 10.1109/BioCAS.2014.6981743

[27] Kim A, Powell CR, Ziaie B. An implantable pressure sensing system with electromechanical interrogation scheme. IEEE Transactions on Biomedical Engineering. 2014; **61**(7):2209-2217

[28] Vaddiraju S, et al. Needle-implantable, wireless biosensor for continuous glucose monitoring. 2015 IEEE 12th International Conference Wearable Implant Body Sensor Networks, BSN 2015. 2015. pp. 2-6. DOI: 10.1109/BSN.2015.7299421

[29] Dehennis A, Mailand M, Grice D, Getzlaff S, Colvin A. A near-field-communication (NFC) enabled wireless fluorimeter for fully implantable biosensing applications. IEEE International Solid-State Circuits Conference. 2013. pp. 298-299. DOI: 10.1109/ISSCC.2013.6487743

[30] Aldaoud A, Laurenson C, Rivet F, Yuce MR, Redouté J. Design of an inductively powered implantable wireless blood pressure sensing interface using capacitive coupling. IEEE/ASME Transactions on Mechatronics. 2015;**20**(1):487-491

[31] Cong P, et al. A wireless and batteryless 130 milligram 300 μW 10-bit implantable blood pressure sensing microsystem for real-time genetically engineered mice monitoring. IEEE International Solid-State Circuits Conference.. 2009;**44**(12):428-429

[32] Park J, et al. A wireless pressure sensor integrated with a biodegradable polymer stent for biomedical applications. Sensors (Switzerland). 2016;**16**(6):1-10. DOI: 10.3390/s16060809

[33] Murphy OH, Bahmanyar MR, Borghi A, McLeod CN, Navaratnarajah M, Yacoub MH, et al. Continuous *in vivo* blood pressure measurements using a fully implantable wireless SAW sensor. Biomedical Microdevices. 2013;**15**(5):737-749

[34] Cheong JH, et al. An inductively powered implantable blood flow sensor microsystem for vascular grafts. IEEE Transactions on Biomedical Engineering. 2012;**59**(9):2466-2475

[35] Gou P, Kraut ND, Feigel IM, Bai H, Morgan GJ, Chen Y, et al. Carbon nanotube chemiresistor for wireless pH sensing. Scientific Reports. 2014;**4**:1-6

[36] Cao H, et al. Batteryless implantable dual-sensor capsule for esophageal reflux monitoring. Gastrointestinal Endoscopy. 2013;**77**(4):649-653

[37] Ativanichayaphong T, Tang SJ, Hsu LC, Huang WD, Seo YS, Tibbals HF, et al. An implantable batteryless wireless impedance sensor for gastroesophageal reflux diagnosis. IEEE MTT-S International Microwave Symposium Digest. 2010. pp. 608-611. DOI: 10.1109/MWSYM.2010.5516775

[38] Farella M, et al. Simultaneous wireless assessment of intra-oral pH and temperature. Journal of Dentistry. 2016;**51**:49-55

[39] Chen LY, et al. Continuous wireless pressure monitoring and mapping with ultra-small passive sensors for health monitoring and critical care. Nature Communications. 2014;**5**:5028

[40] Behfar MH, et al. Biotelemetric wireless intracranial pressure monitoring: An *in vitro* study. International Journal of Antennas and Propagation. 2015 Apr;**2015**:1-10

[41]] Kang S, et al. Bioresorbable silicon electronic sensors for the brain. Nature. 2016;**530** (7588):71-6

[42] Meng X, et al. Dynamic study of wireless intracranial pressure monitoring of rotational head injury in swine model. Electronics Letters. 2012;**48**(7):363

[43] Kneisz L, Unger E, Lanmuller H, Mayr W. *In vitro* testing of an implantable wireless telemetry system for long-term electromyography recordings in large animals. Artificial Organs. 2015;**39**(10):897-902

[44] Zhang Y, et al. A batteryless 19 μW MICS/ISM-band energy harvesting body sensor node SoC for ExG applications. IEEE Journal of Solid-State Circuits. 2013;**48**(1):199-213

[45] Jeon D, et al. An implantable 64nW ECG-monitoring mixed-signal SoC for arrhythmia diagnosis. Dig Tech Papers – IEEE International Solid-State Circuits Conference. 2014;**57**:416-417

[46] Hayami H, et al. Wireless image-data transmission from an implanted image sensor through a living mouse brain by intra body communication. Japanese Journal of Applied Physics. 2016;**55**(4): 04EM03-1 - 04EM03-5. DOI: 10.7567/JJAP.55.04EM03

[47] Anderson GS, Sodini CG. Body coupled communication: The channel and implantable sensors. 2013 IEEE International Conference of Body Sensor Networks, BSN 2013. 2013. pp. 3-7. DOI: 10.1109/BSN.2013.6575490

[48] Enrico Santagati G, Melodia T. Experimental evaluation of impulsive ultrasonic intra-body communications for implantable biomedical devices. IEEE Transactions on Mobile Computing. 2017;**16**(2):367-380

[49] Rasouli M, Phee LSJ. Energy sources and their development for application in medical devices. Expert Review of Medical Devices. 2010;**7**(5):693-709

[50] Gould PA. Complications associated with implantable cardioverter-defibrillator replacement in response to device advisories. Journal of the American Medical Association. 2006 Apr 26;**295**(16):1907

[51] Hudak NS, Amatucci GG. Small-scale energy harvesting through thermoelectric, vibration, and radiofrequency power conversion. Journal of Applied Physics. 2008;**103**(10): 101301-1 - 101301-24. DOI: 10.1063/1.2918987

[52] Mitcheson PD, Yeatman EM, Rao GK, Holmes AS, Green TC. Energy harvesting from human and machine motion for wireless electronic devices. Proceedings of the IEEE. 2008 Sep;**96**(9):1457-1486

[53] Snyder J. Small thermoelectric generators. Electrochemical Society Interface. 2008;**17** (3):54-56

[54] Leonov V, Torfs T, Fiorini P, Van Hoof C. Thermoelectric converters of human warmth for self-powered wireless sensor nodes. IEEE Sensors Journal. 2007;**7**(5):650-656

[55] Bell LE. Cooling, heating, generating power, and recovering waste heat with thermoelectric systems. Science. 2008;**321**(5895):1457-1461

[56] Goto H. Feasibility of using the automatic generating system for quartz watches as a leadless pacemaker power source: A preliminary report. Medical & Biological Engineering & Computing. 1998;**20**(I):9-11

[57] The International Commission on Non-Ionizing Radiation Protection. ICNIRP statement on far infrared radiation exposure. Health Physics. 2006;**91**(6):630-645. link: http://www.icnirp.org/cms/upload/publications/ICNIRPinfrared.pdf

[58] Enwemeka CS. Attenuation and penetration of visible 632.8 nm and invisible infra-red 904 nm light in soft tissues. Official Journal of World Association for Laser Therapy. 2001;**13**:95-101

[59] Amar A Ben, Kouki AB, Cao H. Power approaches for implantable medical devices. Sensors (Switzerland). 2015;**15**(11):28889-28914

[60] Haeberlin A, et al. Successful pacing using a batteryless sunlight-powered pacemaker. Europace. 2014;**16**(10):1534-1539

[61] Katz E, MacVittie K. Implanted biofuel cells operating *in vivo* – methods, applications and perspectives – feature article. Energy & Environmental Science. 2013;**6**(10):2791

[62] Katz E. Implantable biofuel cells operating *in vivo*: Providing sustainable power for bioelectronic devices: From biofuel cells to cyborgs. In: 2015 6th International Workshop on Advances in Sensors and Interfaces (IWASI). IEEE; 2015. pp. 2-13. DOI: 10.1109/IWASI.2015.7184958

[63] Goto K, et al. An implantable power supply with an optically rechargeable lithium battery. IEEE Transactions on Biomedical Engineering. 2001;**48**(7):830-833

[64] Arra S, Leskinen J, Heikkila J, Vanhala J. Ultrasonic power and data link for wireless implantable applications. 2007 2nd International Symposium on Wireless Pervasive Computing. 2007;567-571. DOI: 10.1109/ISWPC.2007.342668

[65] Katz E. Implantable Bioelectronics. 2014. DOI: 10.1002/9783527673148

[66] Kurs A, et al. Wireless power transfer via strongly coupled magnetic resonances. Science. 2007;**317**(5834):83-86

[67] Kim S, Ho JS, Chen LY, Poon ASY. Wireless power transfer to a cardiac implant. Applied Physics Letters. 2012;**101**(7):1-5

[68] Ho JS, Yeh AJ, Neofytou E, Kim S, Tanabe Y, Patlolla B, et al. Wireless power transfer to deep-tissue microimplants. Proceedings of the National Academy of Sciences. 2014;**111**(22):7974-7979

[69] Ho JS, Kim S, Poon ASY. Midfield wireless powering for implantable systems. Proceedings of the IEEE. 2013 Jun 3;**101**(6):1369-1378

[70] Flynn BW, Fotopoulou K. Wireless power transfer in loosely coupled links. Power. 2011;**47**(2):416-430

[71] Flynn BW, Fotopoulou K. Rectifying loose coils: Wireless power transfer in loosely coupled inductive links with lateral and angular misalignment. IEEE Microwave Magazine. 2013;**14**(2):48-54

[72] Aldhaher S, Luk PCK, Whidborne JF. Electronic tuning of misaligned coils in wireless power transfer systems. IEEE Transactions on Power Electronics. 2014;**29**(11):5975-5982

[73] Agrawal DR, et al. Conformal phased surfaces for wireless powering of bioelectronic microdevices. Nature Biomedical Engineering. 2017;**1**(3):43

[74] Visser H. Far-field RF energy transport. IEEE Radio and Wireless Symposium. 2013. pp. 34-6. DOI: 10.1109/RWS.2013.6486632

[75] Mostafalu P, et al. A toolkit of thread-based microfluidics, sensors, and electronics for 3D tissue embedding for medical diagnostics. Microsystems Nanoengineering. 2016;**2**(April):16039

[76] Rogers JA, Someya T, Huang Y. Materials and mechanics for stretchable electronics. Science. 2010;**327**(5973):1603-1607

[77] Yin L, Cheng H, Mao S, Haasch R, Liu Y, Xie X, et al. Dissolvable metals for transient electronics. Advanced Functional Materials. 2014;**24**(5):645-658

[78] Yin L, Huang X, Xu H, Zhang Y, Lam J, Cheng J, et al. Materials, designs, and operational characteristics for fully biodegradable primary batteries. Advanced Materials. 2014;**26**(23):3879-3884

[79] Someya T, Bao Z, Malliaras GG. The rise of plastic bioelectronics. Nature. 2016;**540**(7633):379-385

[80] Hwang S-W, et al. A physically transient form of silicon electronics. Science. 2012; **337**(6102):1640-1644

[81] Luo M, et al. A microfabricated wireless RF pressure sensor made completely of biodegradable materials. Journal of Microelectromechanical Systems. 2014 Feb;**23**(1):4-13

[82] Chavez-Santiago R, et al. Experimental path loss models for in-body communications within 2.36-2.5 GHz. IEEE Journal of Biomedical and Health Informatics. 2015;**19**(3):1-1

[83] Cheffena M. Performance evaluation of wireless body sensors in the presence of slow and fast fading effects. IEEE Sensors Journal. 2015;**15**(10):5518-5526

[84] Ntouni GD, Lioumpas AS, Nikita KS. Reliable and energy-efficient communications for wireless biomedical implant systems. IEEE Journal of Biomedical and Health Informatics. 2014;**18**(6):1848-1856

[85] Ramachandran VRK, Havinga PJM, Meratnia N. HACMAC: A reliable human activity-based medium access control for implantable body sensor networks. BSN 2016 – 13th Annual International Body Sensor Networks Conference. 2016. pp. 383-389. DOI: 10.1109/BSN.2016.7516292

[86] Chavez-Santiago R. Propagation models for IEEE 802.15.6 standardization of implant communication in body area networks. IEEE Communications Magazine. 2013;(August): 80-87. DOI: 10.1109/MCOM.2013.6576343

[87] Santagati GE, Melodia T, Galluccio L, Palazzo S. Medium access control and rate adaptation for ultrasonic intrabody sensor networks. IEEE/ACM Transactions on Networking. 2015;**23**(4):1121-1134

[88] Zheng G, Fang G, Orgun MA, Shankaran R, Dutkiewicz E. Securing wireless medical implants using an ECG-based secret data sharing scheme. 14th International Symposium on Information and Communication Technology – ISC 2014. 2015. pp. 373-377. DOI: 10.1109/ISCIT.2014.7011935

[89] Zheng G, Member S, Fang G, Shankaran R, Orgun MA, Member S, et al. Multiple ECG fiducial points-based random binary sequence generation for securing. IEEE Journal of Biomedical and Health Informatics. 2017;**21**(3):655-663

[90] Pivonka D, Yakovlev A, Poon ASY, Meng T. A mm-sized wirelessly powered and remotely controlled locomotive implant. IEEE Transactions on Biomedical Circuits and Systems. 2012;**6**(6):523-532

[91] Hsieh JY, et al. A remotely-controlled locomotive IC driven by electrolytic bubbles and wireless powering. IEEE Transactions on Biomedical Circuits and Systems. 2014; **8**(6):787-798

Chapter 7

Low-Cost Energy-Efficient Air Quality Monitoring System Using Wireless Sensor Network

Kavi Kumar Khedo and Vishwakarma Chikhooreeah

Additional information is available at the end of the chapter

http://dx.doi.org/10.5772/intechopen.70138

Abstract

Due to rapid industrialization and urbanization, Mauritius is witnessing an unprecedented increase in air pollution. The release of hazardous gases such as carbon monoxide and sulphur dioxide are not only harmful to the health of the population but are also causing irreversible impact to the environment. Currently, there are only two fixed air quality monitoring units on the island and therefore, air pollution cannot be monitored in real-time. The objective of this chapter is to describe the implementation of a low-cost and energy-efficient air quality monitoring system using wireless sensor network (WSN) that can be easily deployed in highly polluted areas of Mauritius. A Hierarchical Based Genetic Algorithm (HBGA) is proposed to address the issue of sensor nodes with limited energy. Based on hierarchical routing and genetic algorithm, HBGA has been designed to extend the lifetime of the network by minimizing the energy consumption. The proposed air quality monitoring system uses an air quality index that can be easily interpreted. The evaluation results confirm the potential of the proposed system for real-time temporal and spatial monitoring of air quality. Moreover, it possible for the general public to have access to the air quality monitoring results in real time.

Keywords: air pollution, sensor networks, genetic algorithms, hierarchical routing, air quality index

1. Introduction

The objective of this chapter is to report on the implementation of a low cost and energy-efficient air quality monitoring system using wireless sensor network (WSN) that is deployed in highly polluted areas in Mauritius. Mauritius is a densely populated island of around 1.3 million people. The air quality in Mauritius has been adversely affected over the past years due to the continuous economic development and alterations in the population consumption and production patterns that have contributed to the release of pollutants in the atmosphere [1]. The increasing level of air pollution in Mauritius from vehicles exhaust fumes, chemical discharge from industries and toxic gas leakages are seriously affecting the citizens' health and damaging crops, ecosystems and materials. The release of gases such as carbon monoxide, sulphur dioxide, oxides of nitrogen, ozone, lead and dust particulates are not only harmful to human health but also have a great impact on the ecosystem [2]. As a result of industrialisation and urbanisation, ensuring a good ambient air quality has become a national challenge in Mauritius.

The emergence of wireless sensor networks (WSNs) has allowed the miniaturization and ubiquity of computing devices. WSNs can be deployed on a global scale for various activities such as environmental monitoring, habitat studying, infrastructure health monitoring, military surveillance, traffic control and others [3]. The construction of smart cities throughout the island can be leveraged to use the WSN medium for more effective and efficient air quality monitoring. There is currently a lack of well-developed network of air quality monitoring systems across the island. The smart city projects in Mauritius is providing the necessary conditions to set up a much better air quality monitoring system in terms of cost, power efficiency, scalability and communication by using cheap portable ambient sensors [4].

The implementation and deployment of the WSN system confirms the potential of such a system to allow the general public to have access to the monitoring results in real time by means of the IoT (Internet of Things) devices. The different WSN components such as sensors, microcontrollers, wireless modules and software that are used to implement the energy-efficient air quality monitoring system are described in this chapter. The system is able to show the real time monitoring readings and graphs of the pollution phenomenon on the monitoring application along with its air quality index (AQI) aspects. The real time readings are made available to any user across the world that is connected to the internet through the ThingSpeak cloud service.

2. Wireless sensor network systems for air quality monitoring

In this section existing works on air quality monitoring and air quality indexing are described. Moreover, existing wireless sensor network (WSN) systems for air quality monitoring are surveyed and discussed.

2.1. Air quality monitoring

Air contains a mixture of gases, small solid and liquid particles. Some substances come from natural sources while others are caused by human activities. The air is said to be polluted when the contents of the air cause harm to the comfort or health of human and animals, or could even damage plants and other materials. These contents are termed as air pollutants and can be either particles, liquids or gaseous in nature. Air Quality Monitoring (AQM) is carried out to assess the extent of pollution, ensure compliance with national legislation, evaluate control options, and provide data for air quality modelling. The goal of AQM is to protect humans and the environment from harmful air pollution [5, 6]. There are different methods to assess any type of pollutant depending on the complexity, consistency and detail of data. These range from simple passive sampling techniques to highly sophisticated remote sensing devices.

The need for the implementation of AQM is to make mitigation strategies and arouse environmental awareness among citizens. Hence, several techniques and technologies have been introduced to monitor air quality [7]. According to a survey done by the World Health Organisation (WHO), it has been seen that urban outdoor air pollution and indoor air pollution accounts for more than 2 million premature deaths each year. More than half of this disease burden is borne by the populations of developing countries. It is therefore vital to constantly monitor the air quality in order to detect unfavourable conditions that must be avoided. Air pollution monitoring consists of systematically assessing the ambient pollutants level in the surrounding indoor and outdoor air. Many countries and cities usually have their own pollution control mechanisms complying with short and long-term air quality objectives set for acceptable levels of pollutants concentrations. Assessing the present and the unforeseen air pollution through continuous air quality monitoring is necessary so as to know the status and trends of ambient air quality and its effects on the environment. Evaluating the changes in air quality is necessary in developing precautionary and corrective measures to control and regulate pollution from various sources.

The increasing level of air pollution is mainly from sources such as smoke from vehicles exhaust and industrial activities. The common gases affecting the quality of air are carbon monoxide, sulphur dioxide, oxides of nitrogen, ozone, lead and dust particulates. Air quality monitoring is therefore needed so that appropriate actions can be taken in order to mitigate its negative impact. Usually databases are used to store the collected data from a monitoring system. The data is then retrieved and analysed to see if they are aligned to the pollution regulatory standards or not. In simple terms air quality monitoring network is used to record the concentration of pollutants and these information are delivered to the population to notify against danger. Another important consideration in air quality monitoring system is the locations of the monitoring stations and networks which should provide proper spatial coverage in populated areas such as busy roads, city centres or a particular location such a hospital and school [8].

Various technology and methodology have been used in order to provide air quality data in real time ranging from traditional way of passive sampling technique to the most sophisticated means such as use of sophisticated remote sensing devices. It is essential to define the options and monitoring methodology most appropriate in terms of cost, reliability and ease of operation. A means of monitoring air pollution is through online General Packet Radio Service (GPRS) sensors comprising of a microcontroller chip and an application server. The mobile data acquisition unit collects the pollution level and organise it into a frame with Global Positioning System (GPS) location, date and time. This frame is then uploaded to the GPRS modem and sent to the pollution server through the public mobile network [5, 9].

Air quality monitoring stations are often expensive and deliver a low resolution sensing data as these stations cannot be densely deployed. Alternatively, one of the effective solutions to provide real time pollution data is through the use of wireless sensor network (WSN) for air quality monitoring which is easy to set up and inexpensive. Consisting of calibrated sensors, WSN systems use a data aggregation algorithm and a routing protocol along with a light-weight middleware for transmission of the pollution data to a base station where they are visualised in graphical forms. Other parameters like humidity and temperature needs to be taken into consideration [9] for providing more accurate pollutant data as these parameters affect the measured gas concentrations.

2.2. Air quality indexing

The assessment and calculation of air pollution level is built on standards which is present in almost every country of the world. The United States Environmental Protection Agency (EPA), the World Health Organization (WHO), the European Commission (EC), the Chinese Ministry of Environmental Protection (MEP) and the Environmental Protecting Department(EPD) of Hong Kong have established different standard limits for pollutants in order to inform the public of the current air quality easily.

A suitable way for characterising atmospheric pollution is through air quality index (AQI). AQI is a quantitative tool which provides information on how fresh or polluted the air is by consolidating the pollution data in the form of reports. Many countries make use of some type of AQI to interpret the quality of the air. An AQI is useful in several ways such as easy interpretation of air quality situation by the general public [10]. Moreover, based on the AQI quick actions can be undertaken, corrective pollution control strategies may be implemented from the trend of events, the impact of regulatory actions may be assessed and scientific researches may be carried out.

The index values help to divide the air pollution situation into categories such that each category is identified by a simple informative descriptor which can be easily used to inform the public on the status of the air as shown in **Table 1** [11].

2.3. Existing air quality monitoring systems

Recent engineering advances together with the internet, communications, and information technologies is enabling the creation of new generation low-cost sensors and actuators that are able to achieve great spatial and temporal resolution and exactitude. A wireless sensor network (WSN) is a wireless network comprising of spatially spread autonomous sensor devices to monitor environmental and physical conditions such as air pollution, light, sound,

Index values	Interval	AQI category
1	$AQI > X_1$	Very good
2	$X_2 < AQI < X_1$	Good
3	$X_3 < AQI < X_2$	Acceptable
4	$X_4 < AQI < X_3$	Poor
5	$AQI < X_4$	Bad

Table 1. Summary of AQI range and descriptor [11].

pressure and temperature etc. The sensors pass their data cooperatively through the network to a central location. There are diverse types of gas sensors that are available to measure different gas concentrations such as CO, CO_2, SO_2, and NO_2 sensors. The WSN system can be deployed in cities to monitor the air pollution level. The air quality measurement is processed and presented to the end user in real time in a user friendly manner. This allows the citizens to take appropriate precautionary measures when required.

The nodes that make up the WSN can range from a few to several hundreds, where each node is connected to its sensor counterpart. Data sensed by each sensor is aggregated by the network and passed on to a sink node. A sensor node may vary in size and typically consist of several parts such as the radio, battery, microcontroller, analogue circuit, and sensor interface. The capabilities of the sensor node are usually constrained in terms of energy, memory, computational speed and communications bandwidth [5, 6]. Nowadays microprocessor developments for WSNs have reduced power consumption with increased processor speed. A WSN system also includes a gateway that provides wireless connectivity back to the distributed nodes through a wireless protocol that depends on the application requirements. Due to its low-power consumption and high-level communication protocols, many WSN systems today are based on IEEE 802.15.4-based specification commonly known as ZigBee. Using WSN for air quality monitoring provides many advantages as listed below:

- Remote and real time measurement monitoring.
- More accurate data for analysis and decision-making.
- Require less human interaction in risky areas.
- Maintain good quality of air and prevent pollution.

In order to overcome the problem of expensive sensing stations for air quality monitoring, researchers from the University of Pisa (UOP) came up with uSense, a sensing system for cooperative air quality monitoring in urban areas [12]. The main advantage of uSense is that it makes use of cheap and small-size sensors that are driven by long-lasting batteries. Moreover the Wi-Fi technology used in uSense allows for fast data transfer such that the air quality information is obtained in real-time. The system has been tested in different areas of interest, monitoring different places with promising results. This system allows its users to know which part of the city has less or no pollution so that they can decide on which routes to take to reach their destination.

In 2008, Ma et al. [13] presented a distributed infrastructure based project, called MoDisNet based on grid computing and WSN technology to monitor air pollution level. In this project, cheap and ubiquitous sensor network is developed to collect large environmental data from road traffic emissions in real-time. The MoDisNet system is based on two layer network framework and a peer-to-peer grid architecture with the implementation of a distributed data mining algorithm to address the challenges of the distributed system. Due to huge amount of data collected and being transferred, research on data fusion and aggregation technique was carried out to improve the system performance. The multi-hop routing capability and grid architecture enabled the development of a data fusion and aggregation technique for the MoDisNet system, thereby saving on the communication cost, reducing the power consumption of the sensor nodes, and increasing the available bandwidth of the wireless channel protocols.

Hu et al. [14] proposed a vehicular wireless sensor network (VSN) architecture to monitor microclimate based on geographic information of vehicles and Global System for Mobile communication (GSM) short messages. One of the objectives of this approach was to demonstrate fine-grained monitoring of the climate using GSM short messages and GPS receivers on vehicles. A prototype was developed using the ZigBee network to monitor the carbon dioxide (CO_2) concentration in specific areas by sending the reported data to a server integrated with Google Maps. In this system, the vehicles are equipped with a CO_2 sensor, a GPS receiver, and a GSM module, forming a ZigBee based intra-vehicle wireless network. Eventually each vehicular sensor travelling inside the area of interest intermittently report their sensed data through GSM short messages.

The SensorWebBike [15] is designed to manage a mobile platform developed with the 'Arduino' open source platform to monitor air quality in cities, in order to integrate the existing monitoring networks and to support public administration in improving urban environment. A circuit board was developed called 'AirQuino', Arduino Shield compatible, integrated with low cost and high resolution sensors, dedicated to the monitoring of environment and air quality, road pavement quality (accelerometer) and the indices of well-being in an urban environment. The board integrates a microprocessor unit that acquires all the sensor readings and analyses fast data from accelerometer and noise sensor. Through General Packet Radio Service (GPRS) technology, the sensor transmits geo-located data on environment and air quality to the server connected to the applications and web server allowing the visualization real time results on a web browser, as shown in **Figure 1**.

The Air Quality Egg [16] is a sensor system designed to collect very high resolution readings of NO_2 and CO concentrations. These two gases are the most indicative elements related to urban air pollution that are sense-able by inexpensive, DIY sensors. There are two versions of the device: an Arduino shield for use by hobbyists, and a more consumer-ready 'hobbyist kit' device. The latter consists of two identical-looking plastic enclosures vaguely resembling white eggs. One unit, the base unit, is connected to the user's Ethernet LAN connection. The second unit monitors NO_2 and CO levels and reports these readings every few minutes back to the base unit via a custom wireless protocol where the readings are sent to Openssensors.io for storage purpose. From there, the data is sent to the Air Quality Egg (AQE) website and to Xively, where the data are transformed into graphs and other visualization. The service also includes the ability to generate triggers for tweets and SMS alerts **Figure 2** shows the Air Quality Egg System diagram.

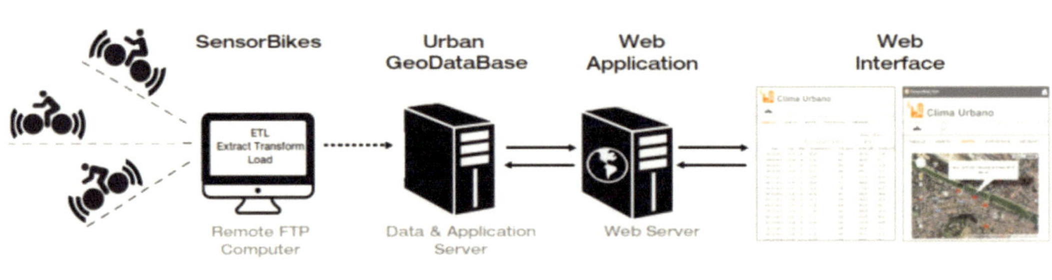

Figure 1. The SensorWebBike framework components [15].

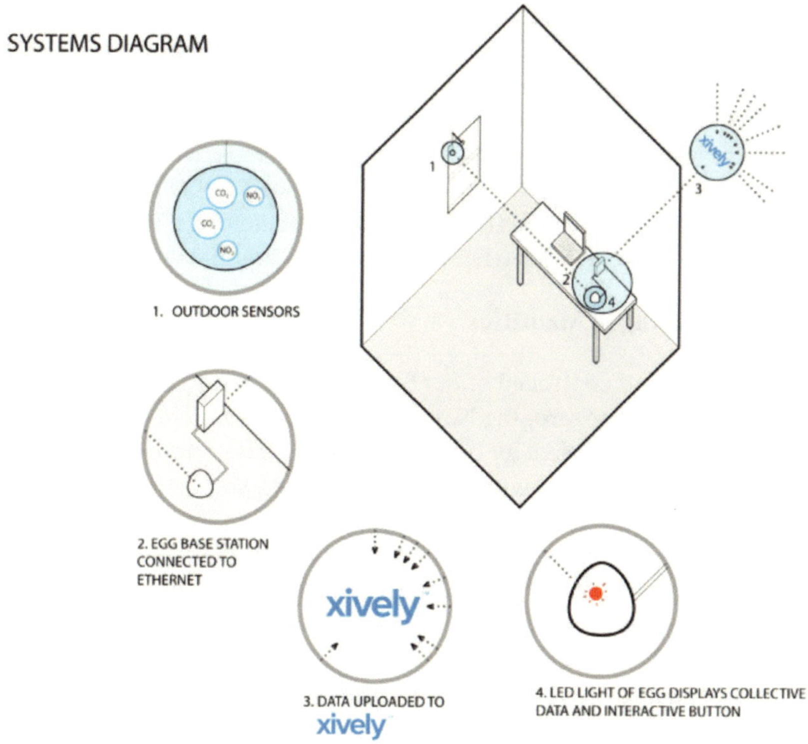

Figure 2. Air Quality Egg system diagram [16].

3. Situational analysis

This section describes the current air pollution problems and existing measures being undertaken to alleviate those problems in the Mauritian context by the local authorities. The related constraints pertaining to such a situation in Mauritius are analyzed.

3.1. Air quality in Mauritius

Clean fresh air is vital for human well-being and good health. However in Mauritius, an increase in air pollution level has been observed over the past few years due to industrialization and urbanization in the country whereby maintaining a good ambient air quality has become a challenge. Moreover drastic changes in the population production and consumption behaviour as well as continuous economic development have contributed to the rise of air pollution in Mauritius.

Some of the most common pollutants in Mauritius include carbon monoxide, sulphur dioxide, nitrogen oxide, ozone and particulate matter. The sources of air pollution in Mauritius are from industrial activities, electricity generation, transportation and as well as from burning of solid waste. Furthermore it is in the regions termed as industrial hotspots such as Valentina - Phoenix,

Cité St. Luc - Forest Side, Terre Rouge, La Tour Koenig and Cité Vallejee where higher levels of air pollution have been recorded [17]. Also the exhaust emission from the vehicles is one of the major contributors of urban air pollution in Mauritius. There are also some cases of improper medical waste incineration releasing harmful pollutants and injurious metals such as lead and mercury into the environment affecting the inhabitants and crops. The health effects linked with the unsafe pollutants range from mild irritations of eye, nose, throat and skin to more serious diseases like asthma, chronic bronchitis and lung cancer. Therefore, it is observed that air pollution is having adverse impact on human health, crops, materials and the environment in general.

3.2. Air quality monitoring in Mauritius

Air pollution in Mauritius is controlled under the 1998 Environment Protection (Standards for Air) Regulations and in this regard, the National Environmental Laboratory (NEL) had been set up for monitoring of the ambient air quality in public places, industrial zones and residential areas. Air quality is monitored with a fixed monitoring station dispatched at Cassis and a mobile station used in various regions of the island where a high level of pollution is suspected. Special equipment at the stations are used to measure the concentrations of major pollutants such as carbon monoxide, sulphur dioxide, nitrogen oxide, and ozone [1]. Recently two more fixed ambient air monitoring stations were acquired by the Ministry of Environment and are in operation at the Mauritius Meteorological Services, Vacoas, and at the Islamic Cultural Centre, Port Louis, respectively. Data obtained from these stations are eventually interpreted by the NEL officers to ensure its compliance against the air quality standards.

Several steps have been undertaken to ensure better air quality and reduction of atmospheric pollution by the government of Mauritius. For example new air quality standards have been set, unleaded petrol has been introduced and chlorofluorocarbons (CFCs) have been phased out. Also concerning transport sector, monitoring of vehicular emissions as per the Road Traffic Regulations 2002 is carried by the National Transport Authority assisted by the environment police for visual road-side checks [1]. Moreover, with the ambitious economic development programme undertaken by the government of Mauritius regarding construction of smart cities throughout the island, there is a need to have environment monitoring system in place aligned with government vision of having technology driven facilities for creating a pleasant and clean living environment free from any pollution and nuisances.

3.3. Problem analysis

As Mauritius is becoming a more industrialized, urbanized and densely populated nation, maintaining a good ambient air quality becomes a major challenge. It is recognized that there is a pressing need to address the air pollution issues in Mauritius. With only a few air quality monitoring stations available, systematic monitoring of ambient air quality across the island is very difficult. Moreover due to lack of a well-built network, tracking the evolution of air quality in several locations simultaneously is presently not possible [17].

The conventional air pollution monitoring system used in Mauritius have limited scalability and besides being bulky and expensive, these monitoring stations require trained technical staff to be operated and requires continuous maintenance, upgrade and repairs to prevent reduction in their lifespan. Furthermore the sites of the monitoring stations necessitate careful placement to be effective as frequent relocation of the stations can be costly and time consuming. It is therefore difficult for the authorities to accomplish effective and efficient air quality monitoring exercises across the island.

Indeed, the conventional monitoring system is bulky and expensive necessitating a lot power for its operation. Therefore in order to overcome the difficulties and high costs involved in monitoring air quality using wired devices, a new wireless air monitoring system is required that is less costly and more power efficient with low energy consumption. Moreover due to the communication constraints of the conventional monitoring system, it cannot cover large-scale areas for monitoring. As such, there is a need to have a reliable and fault tolerant communication with minimum consumption of energy in order to have an enhanced air quality monitoring (AQM) system.

The AQM system should have the capability to inform the citizens about the concentrations of different gases in the air and create environmental consciousness so as to minimize the degradation of air quality. The use of sensor networks represents a solution for strong environmental surveillance. Data from sensor networks can capture more accurate input conditions which result in more reliable conclusion about air quality. The deployment of sensor networks can also contribute in air emission inventories and detecting pollution hotspots, as well as allowing real-time exposure assessment for deriving abatement strategies. Data retrieval from sensors is much direct and straightforward. Therefore, the use of sensor networks for AQM provides granularity that better identifies the presence of pollution sources and helps in the studies on the effects of air pollution on socio-ecological justice and human quality of life [18].

4. A low-cost energy-efficient WSN system for air quality monitoring

The air quality monitoring system proposed collects the ambient pollutant gases such as carbon monoxide (CO), ozone (O_3), nitrogen dioxide (NO_2) and sulphur dioxide (SO_2). The gas sensors nodes deployed outdoor in the region of interest are classified into several clusters where each cluster consists of one cluster head and several member nodes. The member nodes send their data to the cluster heads (CHs) for their respective clusters. The CHs then forwards the aggregated data to the sink node. The data from the WSN is sent over to the gateway which forwards them to a cloud server where the data is stored and processed into graphical visualizations for the user's monitoring purposes. The high level structure of the proposed system is shown in **Figure 3**.

4.1. Proposed algorithm for energy efficient hierarchical clusters generation

A genetic algorithm (GA) is an efficient algorithm that imitates the processes of biological evolution for solving particular problems. The idea behind this algorithm is to create a new population that is better and fitter from the existing population. The three main operations that are normally carried in the genetic algorithm include selection, crossover and mutation [19]. Each of the operations is briefly described below.

1. **Selection:** Selection is generally the first operator applied for selecting the fitter chromosomes from the population to be used afterwards in the crossover process.
2. **Crossover:** In this method two chromosomes are combined to produce a new offspring which is expected to be better than its parents provided it retains the best characteristics of both parents.
3. **Mutation:** After a crossover is performed, mutation takes place for maintaining genetic diversity among the generations of the chromosomes population. This technique changes the gene values in a chromosome to a different state enabling the genetic algorithm to arrive at better solution than was previously possible.

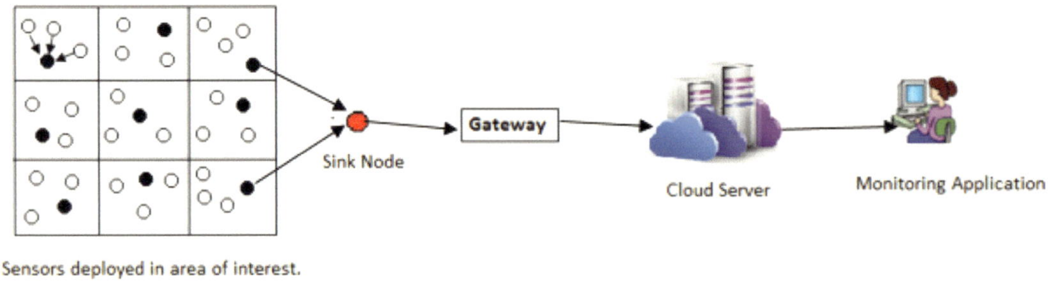

Figure 3. High level architecture of the proposed system.

4.1.1. Hierarchical Based Genetic Algorithm

A Hierarchical Based Genetic Algorithm (HBGA) is proposed to minimize the communication consumption energy of all sensor nodes and to efficiently maximize the network lifetime by finding the optimum number of CHs. In the proposed genetic algorithm, the nodes are represented as bits of a chromosome where the cluster head nodes are represented as 1s and the member nodes as 0s. Several chromosomes make up a population from which the best chromosome is used to create the next population. The parameters such as current energy levels and transfer distances are used to determine the fitness of the chromosome so that the population is transformed into a new generation based on its survival fitness.

- Proposed selection method

The method to be used for selecting chromosomes for parents to crossover shall be the tournament selection method in which several 'tournaments' are run among a few chromosomes chosen at random from the population. The winner of each tournament is the one with the best fitness which is eventually selected for crossover.

- Proposed crossover method

The crossover operator to be applied in the proposed algorithm is the uniform crossover technique. Using a probability ratio known as the mixing ratio, the uniform crossover operator decides which parent will contribute how much gene values in the offspring chromosomes. Consider the two parents selected for crossover as shown in **Figure 4**.

Parent 1	1	1	0	1	1	0	0	1	0	0	1	1	0	1	1	0
Parent 2	1	1	0	1	1	1	1	0	0	0	0	1	1	1	1	0

Figure 4. Crossover parent [19].

If the mixing ratio is set to 0.5, then half of the genes in the offspring will come from parent 1 and other half will come from parent 2 as shown in **Figure 5**.

- Proposed mutation method

The mutation operator to be used in the proposed algorithm is the Flip Bit method which basically inverts the value of the chosen gene. For example 2 original off-springs are selected for mutation as shown in **Figure 6**.

The value of the chosen gene is inverted from 0 to 1 and 1 to 0 and the mutated offspring produced are: Mutated offspring 1 Mutated offspring 2 as shown in **Figure 7**.

4.2. The proposed air quality monitoring system

The proposed AQM system uses a wireless sensor network with low-cost sensors and hardware components along with the necessary software to effective monitor the air pollution phenomenon. The WSN components used for the proposed system consist of microcontrollers, wireless modules and different gas sensors. The system uses the MQ series semiconductor gas sensors. Each MQ gas sensor is sensitive to a specific gas such as carbon monoxide, nitrogen dioxide, butane, hydrogen, and smoke among others that use a small heater with an electrochemical sensor. These sensors can be calibrated using the load-resistor and burn-in technique. Its output is read with an analogue input of the Arduino platform. **Table 2** provides a list of gas sensors that is used in the system implementation.

The monitoring node consists of an Arduino Uno microcontroller board built on the ATmega328. It has 14 digital input/output pins with 6 analogue inputs, a 16 MHz ceramic resonator and a USB connection. The ATmega328 has a pre-burned bootloader that permits upload of new code to it without the need of an external hardware programmer. The program inside this chip coordinates the data transmission to the central server over an UART TTL (5 V) serial communication.

The radio frequency (RF) module comprises of an RF Transmitter and an RF Receiver. The RF 433 MHz modules have a wide applications in various systems that require wireless control. These low-cost modules can be used with any microcontroller. The transmitter/receiver (Tx/Rx) pair functions at a frequency of 433 MHz. The RF transmitter receives serial data from the

Offspring 1	1_1	1_2	0_2	1_1	1_1	1_2	1_2	0_2	0_1	0_1	0_2	1_1	1_2	1_1	1_1	0_2
Offspring 2	1_2	1_1	0_1	1_2	1_2	0_1	0_1	1_1	0_2	0_2	1_1	1_2	0_1	1_2	1_2	0_1

Figure 5. Crossover offspring [19].

Original offspring 1	1	1	0	1	1	1	1	0	0	0	0	1	1	1	1	0
Original offspring 2	1	1	0	1	1	0	0	1	0	0	1	1	0	1	1	0

Figure 6. Original offspring [19].

Mutated offspring 1	1	1	0	0	1	1	1	0	0	0	0	1	1	1	1	0
Mutated offspring 2	1	1	0	1	1	0	1	1	0	0	1	1	0	1	0	0

Figure 7. Mutated offspring [19].

sensor and transmits it wirelessly using radio frequency to the receiver through its antenna at the rate of 1–10 Kbps. A gateway application is implemented to read incoming data from the receiver module through serial port communication and sending the data to the ThingSpeak cloud server.

The sensing node is set up such as the pollutant measurements are read from the analogue output of the microcontroller. The MQ-7 sensor is wired to the first Arduino board together with the transmitter as shown in **Figure 8**. The receiver is connected to the second Arduino board as shown in **Figure 9**. The MQ-7 sensor would require calibration of about 30 s sensor heating cycle and a 60 s sampling cycle before the actual pollutant concentration values transmitted can be considered.

The AQI breakpoints shown in **Table 3** are used for AQI calculation [20] in the air quality monitoring system. This particular AQI breakpoints is used as it provides a simple and easily interpretable air quality condition that may be used by the general public.

The AQI is calculated using the recorded pollutant concentration data with the following equation [20]:

$$I_p = \frac{I_{Hi} - I_{Lo}}{BP_{Hi} - BP_{Lo}} (C_p - BP_{Lo}) + I_{Lo} \quad (1)$$

Where I_p = the index for pollutant p; C_p = the rounded concentration of pollutant p; BP_{Hi} = the breakpoint that is greater than or equal to C_p; BP_{Lo} = the breakpoint that is less than or equal to C_p; BP_{Hi} = the breakpoint that is greater than or equal to C_p; I_{Hi} = the AQI value corresponding to BP_{Hi}; I_{Lo} = the AQI value corresponding to BP_{Lo}.

For the purpose of the air quality monitoring system, the ThingSpeak Internet of Things (IoT) cloud service provider is used to store the sensed data. The ThingSpeak service makes provision for sensed data to be uploaded to its server using IoT devices such as Arduino and Raspberry Pi. By using the ThingSpeak's write Application Developer Interface (API) key, the sensed data from the sensor nodes are sent to the cloud server. The data are then fetched, analysed and visualized in the client monitoring application in real time as shown in **Figure 10**.

Gas sensor ID	Monitoring gas
MQ-131	Ozone
MQ-7	Carbon monoxide
MiCS-2714	Nitrogen dioxide
MQ-136	Sulphur dioxide

Table 2. MQ series semiconductor gas sensors.

Figure 8. Transmitter setup.

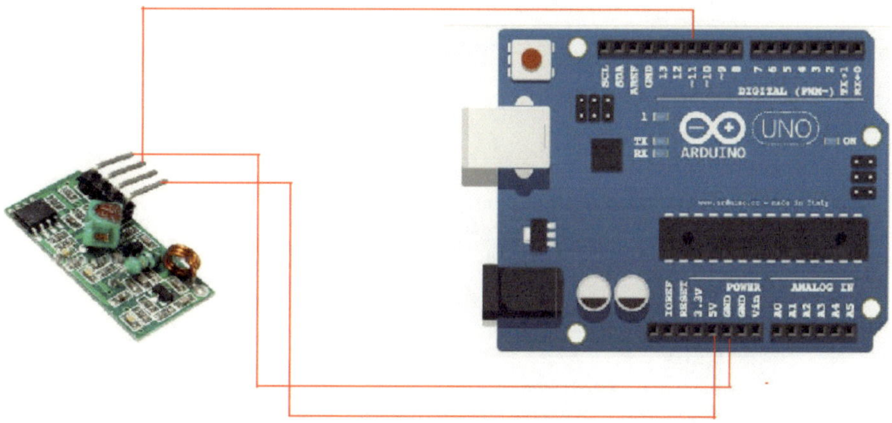

Figure 9. Receiver setup.

The monitoring application allows the visualization of the sensors deployed in different regions as shown in **Figure 11**. Moreover, it provides the end user with graphical representation of the current level of gas concentration detected in real time. The monitoring application will also enable the user to view the current AQI value with respect to the gas concentration measured as shown in **Figure 12**.

O_3 (ppm) 8-hour	CO (ppm)	SO_2 (ppm)	NO_2 (ppm)	AQI	Category	Colour code
0.000–0.064	0.0–4.4	0.000–0.034	*2	0–50	Good	Green
0.065–0.084	4.5–9.4	0.035–0.144	*2	51–100	Moderate	Yellow
0.085–0.104	9.5–12.4	0.145–0.224	*2	101–150	Unhealthy for Sensitive Groups	Orange
0.105–0.124	12.5–15.4	0.225–0.304	*2	151–200	Unhealthy	Pink
0.125–0.374	15.5–30.4	0.305–0.604	0.65–1.24	201–300	Very Unhealthy	Red
*1	30.5–40.4	0.605–0.804	1.25–1.64	301–400	Hazardous	Grey
*1	40.5–50.4	0.805–1.004	1.65–2.04	401–500	Extremely Hazardous	Black

*1—8-hour O_3 values do not define higher AQI values (>301).
*2—NO_2 can generate an AQI only above a value of 200.

Table 3. AQI breakpoints.

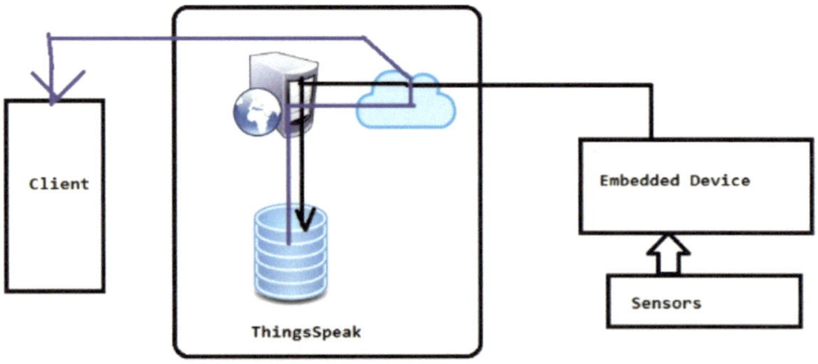

Figure 10. The service architecture of ThingSpeak.

Figure 11. Monitoring application interface.

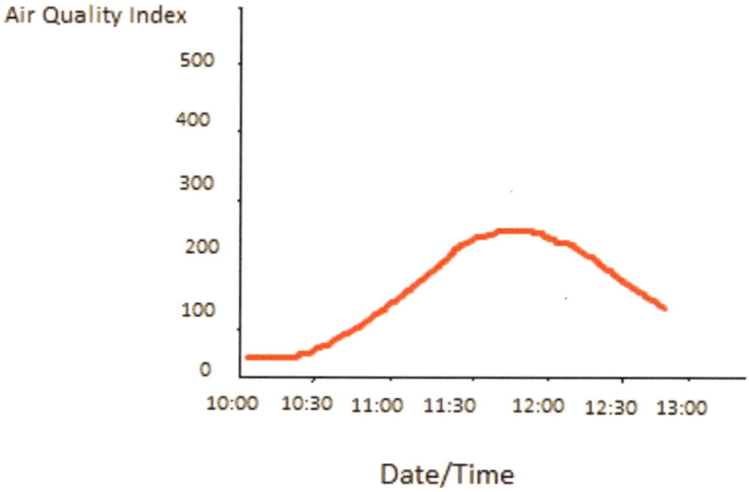

Figure 12. Graphical display of AQI value.

5. Results and discussions

A number of tests were carried out to assess the performance of the proposed HBGA technique. The results are compared with the popular LEACH protocol with varying simulation parameters to determine its efficiency for several scenarios. The input parameters for the tests include the number of nodes, the sensor network area covered in metres square and the number of rounds. The results of the tests include the dissipation energy, the remaining energy and the number of dead nodes of the network for each round. Also the overall network lifetime is derived from the remaining energy at the end of each test (**Table 4**).

Table 5 summarizes the list of tests carried out. Each scenario is different from one another in terms of the constant and varying parameters used.

Parameter	Type	Description
Number of nodes	Input	Number of nodes indicates the size of the network. The algorithm is tested with varying number of nodes to study its scalability
Area	Input	Network coverage area in metres square
Number of rounds	Input	Number of iteration for cluster formation
Dissipation energy	Output	Energy capacity loss during clustering and transmission
Remaining energy	Output	Energy capacity left of the network
Number of dead nodes	Output	Indicates the number of dead nodes for the network
Overall network lifetime	Output	Sum of energy remaining for each round

Table 4. List of test parameters.

Scenario	Varying parameter	Constant parameter	Output
1	Number of nodes	Area Number of rounds CH probability	Residual energy against number of rounds
2	Area	Number of nodes Number of rounds CH probability	Residual energy against number of rounds
3	Number of rounds	Number of nodes Area CH probability	Residual energy against number of rounds
4	Number of nodes	Area Number of rounds CH probability	Remaining energy against number of rounds
5	Area	Number of nodes Number of rounds CH probability	Remaining energy against number of rounds
6	Number of rounds	Number of nodes Area CH probability	Remaining energy against number of rounds
7	Number of nodes	Area Number of rounds CH probability	Number of dead nodes against number of rounds
8	Area	Number of nodes Number of rounds CH probability	Number of dead nodes against number of rounds
9	Number of rounds	Number of nodes Area CH probability	Number of dead nodes against number of rounds

Table 5. List of test scenarios.

5.1. Dissipation energy

From the results obtained in scenarios 1, 2 and 3, it is observed that the dissipation energy values for the HBGA technique do not fluctuate much compared to the LEACH protocol. It is seen that there is higher energy dissipation during the first 40 rounds using the LEACH protocol compared to the HBGA technique in which the dissipation energy values remain more or less the same throughout all the rounds. The LEACH protocol has lower energy dissipation after about 40 rounds due to the fact that the remaining alive nodes are located nearer to the BS. Therefore in the LEACH protocol much energy is dissipated for the farthest nodes from the BS making them to die faster compared to the HBGA in which the dissipated energy remains consistent for longer number of rounds.

5.2. Remaining energy

From the results obtained in scenarios 4, 5 and 6, it observed that as the number of rounds increases, the remaining energy decreases for both HBGA and LEACH protocol as expected. However there is much difference in the decreasing remaining energy values between the two techniques. In HBGA the remaining energy for the network remains much higher than that of the LEACH protocol for the same number of rounds and other parameters. For example in scenario 4 for increasing number of nodes, the remaining energy from the HBGA technique remains approximately 2 times higher than that of the LEACH protocol for the last round. Similarly in scenario 5 for increasing area, the remaining energy from the HBGA technique is noticed to be approximately 3 to 6 times higher than that of the LEACH protocol for the last round. Likewise the remaining energy from the HBGA technique is observed to be approximately 2–3 times higher than that of the LEACH protocol for the last round in the 6th scenario for increasing number of rounds. It can therefore be concluded that HBGA is more energy-efficient than the popular LEACH protocol.

5.3. Dead nodes

From the results obtained in scenarios 7, 8 and 9, it is found that as the number of rounds increases, the number of dead nodes also increases for both HBGA and LEACH protocol which is expected due to energy loss. However there is a huge difference in the number of dead nodes between HBGA and LEACH for the same number of rounds. For example in scenario 7 where number of nodes is 50, it can be seen that at the end of simulation only 2 nodes are dead using the HBGA technique compared to 22 dead nodes using the LEACH protocol. From scenario 8 it is observed that the number of dead nodes increases as the network area increases. In this scenario as well a large difference in the number of dead nodes is found between HBGA and LEACH. For example in a network area of $600 \times 600\ m^2$, 22 dead nodes is observed using the HBGA technique and 80 dead nodes is observed using the LEACH protocol. Also a common observation from scenarios 7, 8 and 9 is that in HBGA, the number of dead nodes remains 0 for a higher number of rounds compared to LEACH.

5.4. Overall system test

Moreover, it is observed that the network lifetime of HBGA is always greater than that of the LEACH protocol for all the scenarios tested whether it is varying nodes, varying distance or varying rounds. The HBGA performs better than LEACH in all of the scenarios because HBGA make use of the hierarchical cluster based routing and genetic algorithm technique making the network survive for a longer time. The proposed HBGA technique has proven to be a better technique than the LEACH protocol and has been able to meet its requirements in terms of energy efficiency, robustness, prolonged network lifetime and scalability.

The air quality monitoring (AQM) system has been tested for different air quality conditions and it has shown its effectiveness by displaying the real data successfully on the monitoring application. Moreover, the corresponding AQI is displayed for the different air quality situation. This proves the effectiveness of the system in successfully showing the air pollution level in real time by means of the RF wireless transmission of the data to the receiver from where it is eventually sent to the cloud server.

6. Conclusion

Mauritius is witnessing an unprecedented level of air pollution lately as attested by the latest statistics report (Statistics Mauritius—environment statistics year 2014, 2015). In order to ensure air quality standards are being respected, an effective and efficient monitoring system is required. In this chapter the implementation of a low-cost efficient air monitoring system using the wireless sensor network is described. WSNs consisting of low-priced components such as tiny sensor nodes, microcontrollers and wireless modules have proven their effectiveness and efficiency in many areas. The main challenge in the design of communication protocols for wireless sensor network is energy efficiency due to the limited amount of energy in the sensor nodes.

A Hierarchical Based Genetic Algorithm (HBGA) has been put forward to achieve energy-efficient cluster formation and minimal energy dissipation of the nodes resulting in an extended network lifetime. The testing results shown that the HBGA scheme offers a better performance than the LEACH protocol in terms of energy consumption and network lifetime. Therefore, the proposed clustering approach is more energy efficient and hence effective in prolonging the network life time compared to the LEACH protocol. The proposed system has proven air quality monitoring using WSN to be effective and achievable. The working system further confirms the potential of making it possible for the general public to have access to the air quality monitoring results in real time by means of IoT devices connected to the internet. The air quality system can be enhanced to include a module for predicting the air pollution propagation in cases of gas leakages so that the people can be evacuated safely before the occurrence of any serious threats to their health. The pollution propagation can be predicted based on the gas diffusion theory or by using a diffusion model in which the windy conditions can be used to locate the pollution source and forecast its dispersion pattern.

Author details

Kavi Kumar Khedo* and Vishwakarma Chikhooreeah

*Address all correspondence to: k.khedo@uom.ac.mu

Faculty of Information, Communication and Digital Technologies, University of Mauritius, Reduit, Mauritius

References

[1] Ministry of Environment and Sustainable Development (MOESD). Mauritius Environment Outlook Report 2011 [Internet]. 2011. Available from: http://environment.govmu.org/English/DOCUMENTS/MAURITIUS%20ENVIRONMENT%20OUTLOOK%20REPORT%20SUMMARY%20FOR%20DECISION%20MAKERS.PDF [Accessed: 2017/05/10]

[2] Hoek G, Krishnan RM, Beelen R, Peters A, Ostro B, Brunekreef B, Kaufman JD. Long-term air pollution exposure and cardio-respiratory mortality: A review. Environmental Health. 2013 May 28;**12**(1):43

[3] Mead MI, Popoola OA, Stewart GB, Landshoff P, Calleja M, Hayes M, Baldovi JJ, McLeod MW, Hodgson TF, Dicks J, Lewis A. The use of electrochemical sensors for monitoring urban air quality in low-cost, high-density networks. Atmospheric Environment. 2013 May 31;**70**:186-203

[4] Khedo KK, Perseedoss R, Mungur A. A wireless sensor network air pollution monitoring system. arXiv preprint arXiv:1005.1737. 2010 May 11

[5] Al-Ali AR, Zualkernan I, Aloul F. A mobile GPRS-sensors array for air pollution monitoring. IEEE Sensors Journal. 2010 Oct;**10**(10):1666-1671

[6] Postolache OA, Pereira JD, Girao PS. Smart sensors network for air quality monitoring applications. IEEE Transactions on Instrumentation and Measurement. 2009; **58**(9):3253-3262

[7] Penza M, Suriano D, Villani MG, Spinelle L, Gerboles M. Towards air quality indices in smart cities by calibrated low-cost sensors applied to networks. IEEE SENSORS 2014 Proc. 2012-2017. DOI: 10.1109/ICSENS.2014.6985429

[8] Popescu F, Ionel N, Lontis L, Calin I, Dungan I. Air quality monitoring in an urban agglomeration. Optoelectronic Techniques for Environmental Monitoring. 2009;**56**:495-506

[9] Sirisikar S, Karemore P. Review paper on air pollution monitoring system. International Journal of Advanced Research in Computer and Communication Engineering. 2015 Jan;**4**(2):218-220

[10] Sharma M, Bhattacharya A. National Air Quality Index – India. Central Pollution Control Board [Internet]. 2014. p. 5-8. Available from: http://www.indiaenvironmentportal.org.in/files/file/Air%20Quality%20Index.pdf [Accessed: 2017/05/10]

[11] Taieb D, Brahim AB. Methodology for developing an air quality index (AQI) for Tunisia. International Journal of Renewable Energy Technology. 2013;**4**(1):86. DOI: 10.1504/ijret.2013.051067

[12] Brienza S, Galli A, Anastasi G, Bruschi P. A low-cost sensing system for cooperative air quality monitoring in urban areas. Sensors. 2015;**15**(6):12242-12259. DOI: 10.3390/s150612242

[13] Ma Y, Richards M, Ghanem M, Guo Y, Hassard J. Air pollution monitoring and mining based on sensor grid in London. Sensors. 2008;**8**(6):3601-3623. DOI: 10.3390/s8063601

[14] Hu SC, Wang YC, Huang CY, Tseng YC. Measuring air quality in city areas by vehicular wireless sensor networks. Journal of Systems and Software. 2011;**84**(11):2005-2012. DOI: 10.1016/j.jss.2011.06.043

[15] Vagnoli C, Martelli F, De Filippis T, Di Lonardo S, Gioli B, Gualtieri G, Matese A, Rocchi L, Toscano P, Zaldei A. The SensorWebBike for Air Quality Monitoring in a Smart City. IET Conference on Future Intelligent Cities, 4-5 Dec. 2014, London, UK

[16] Wicked Device. Air Quality Egg [Internet]. Available from: https://airqualityegg.wicked-device.com/ [Accessed: 2017/04/25]

[17] Acclimate. Mauritius Environmental Report [Internet]. 2011. Available from: http://acclimate-oi.net/files/documentation/20110627_mauritius-environmental-report.pdf? [Accessed: 2017/02/15]

[18] Kumar P, Morawska L, Martani C, Biskos G, Neophytou M, Di Sabatino S, Bell M, Norford L, Britter R. The rise of low cost sensing for managing air pollution in cities. Environment International. 2015;**75**:199-205

[19] Chakraborty RC. Fundamentals of genetic algorithms. Reproduction. 2010 Jun;**22**:35

[20] Mintz D. Guidelines for the Reporting of Daily Air Quality—The Air Quality Index (AQI). Vol. 1. North Carolina: U.S. Environmental Protection Agency Research Triangle Park; 2006. pp. 12-16

Chapter 8

Modern Clustering Techniques in Wireless Sensor Networks

I.S. Akila, S.V. Manisekaran and R. Venkatesan

Additional information is available at the end of the chapter

http://dx.doi.org/10.5772/intechopen.70382

Abstract

Wireless sensor networks (WSNs) are employed in various applications from healthcare to military. Due to their limited, tiny power sources, energy becomes the most precious resource for sensor nodes in such networks. To optimize the usage of energy resources, researchers have proposed several ideas from diversified angles. Clustering of nodes plays an important role in conserving energy of WSNs. Clustering approaches focus on resolving the conflicts arising in effective data transmission. In this chapter, we have outlined a few modern energy-efficient clustering approaches to improve the lifetime of WSNs. The proposed clustering methods are: (i) fuzzy-logic-based cluster head election, (ii) efficient sleep duty cycle for sensor nodes, (iii) hierarchical clustering, and (iv) estimated energy harvesting. Classical clustering approaches such as low energy adaptive clustering hierarchy (LEACH) and selected contemporary clustering methods are considered for comparing the performance of proposed approaches. The proposed modern clustering approaches exhibit better lifetime compared to the selected benchmarked protocols.

Keywords: wireless sensor networks, clustering, energy, harvesting, sleep duty cycle, fuzzy logic

1. Introduction

1.1. Wireless sensor networks

In the recent years, wireless sensor networks (WSNs) have attained significant roles in various applications and have attracted the attention of researchers due to their complex, multifaceted requirements which often divulge inherent tradeoffs. A wireless sensor network is made up of sensor nodes connected through an ad hoc and self-configuring connectivity.

A wireless sensor network consists of a set of spatially distributed autonomous sensor nodes to monitor environmental parameters such as temperature, pressure, etc., and to cooperatively pass their data through the network to a primary location called sink. The sink of a WSN collects data from the sensor nodes and reports the same to users through the Internet or through any private virtual network (PVN). By nature, WSNs inherit features and requirements of an ad hoc network. WSN belongs to the class of low range wireless personal area network (LPWPAN).

A sensor node consists of a radio transceiver (which performs the role of both transmitter and receiver), a microcontroller, and an electronic circuit for interfacing with the associated sensors and an energy source (usually a battery or an embedded form of energy harvesting). The cost and size of sensor nodes show a significant degree of variation depending upon the nature of the applications. In many applications, sensor nodes demand self-organization due to the randomness present in uncontrolled, non-deterministic topologies.

Depending upon the nature of applications, various categories of sensor nodes are provided for monitoring parameters such as temperature, moisture, sound, motion of objects, etc. In essence, sensor networks compensate for human efforts in inaccessible terrains and present more comfortable, smart snapshots of the environment. In the recent future, sensor networks would conquer an integral part of human life and make existing personal computers, mobile communication devices and other computing devices less popular.

A sensor network may be composed of homogenous or heterogeneous sensor nodes. They may monitor either space or objects or interactions of these two. Today, sensor networks are employed in diversified fields such as battle field surveillance, medical diagnostics, precision agriculture, weather monitoring and home appliances control. Every sensor application demands its own set of requirements and characteristics. Some sensor applications employ reactors in the place of ordinary sensors to react to the events in an appropriate manner.

Design of WSNs exhibits challenges due to the limited resources in terms of storage, processing and communication of messages. In most of the sensor applications, these resources become non-renewable also. Theoretical estimation could not be accurate enough in many scenarios to predict and prevent failure of sensor networks. The design complexity of WSNs increases with emerging applications and their requirements. Conventional algorithms designed for ad hoc networks are not good enough to cater to the needs of these sensor applications and this mandates new policies and protocols to be evolved.

Wireless sensor networks can be classified into pre-deterministic and unattended networks based on the type of applications. Former category of networks gains the advantages of quality of service (QoS), fault tolerance, robustness and scalability. In many pragmatic scenarios, human supervision for sensor networks is limited or prohibited since the nodes are dispersed in critical environments such as deeper part of jungles and underwater environments. These networks are called as unattended networks.

Also, an inherent trade-off is observed amidst the parameters used to determine the performance of a WSN. The applications seldom reconcile with identical set of parameters. Juxtaposing the requirements, they resist generalized solutions owing to their nature of self-contradiction and application-specific precincts. As the autonomy of nodes increase, scalability of the solutions is challenged. Research efforts made to improve throughput often result in increased overhead. There is a combined need for fast convergence time and minimum energy consumption of sensor nodes. When the solutions are inclined toward one or a set of parameters, they habitually compromise the rest of the performance factors. This intricacy leads to many interesting queries and solutions in describing the efficiency of a WSN. Slicing over the temporal and spatial domains, the process becomes more complex, multifaceted and highly specialized.

The lifetime of a sensor network is stanchly dependent on the energy consumption, especially when there is no provision for human access to the involved sensor nodes. Hence, many methods have been proposed to minimize energy consumption in wireless sensor networks. The design of wireless sensor networks exhibits many challenges from this perspective.

The performance of a wireless sensor network is not only multifaceted but also inherently imbalanced under one or limited angles of perception. A holistic and fair approach requires an unambiguous and complete understanding of sensor applications.

1.2. Clustering in wireless sensor networks

Sensor nodes in an environment collect data and transmit it to a sink either directly or collaboratively through other nodes. Many sensor applications cluster the sensor nodes to achieve scalability, robustness and reduced network traffic.

A sample scenario of clustering is shown in **Figure 1**. Here, clusters are provided with cluster heads and these cluster heads transmit the aggregated data to the base station or the sink.

The primary advantage of clustering is the scalability of performance across the expanding sensor networks. In addition to this, clustering approach provides numerous secondary advantages. It ensures reliability and avoids one-point failure due to its localized solutions. A clustering solution can suggest a sleep/wakeup schedule for a WSN to effectively reduce power consumption. In many sensor applications, all the sensor nodes are not required to be in wakeup state and consume energy. Based on the temporal and spatial dependencies, some sensor nodes can be put in sleep mode in which no energy is consumed. An effective schedule can be devised and communicated to these sensor nodes through the sink or administrator. Also, clustering ensures scalability of the application performance due to its semi-distributed nature.

As indicated in the work done by Abbasi and Younis [1], clustering possesses certain challenges amidst its advantages. There are selected sensor nodes identified as "cluster heads," which monitor and regulate the data flow across clusters for which considerable energy is consumed. Hence, the process of reclustering and reelection of cluster heads is required which results in the reduced lifetime of sensor networks.

One of the conventional clustering protocols named low energy adaptive clustering hierarchy (LEACH) [2] addresses the overloading of clusters and it rotates the role of cluster heads among the sensor nodes present in a cluster. The significant drawback of this approach is that no weightage is given for the residual energy of the sensor nodes. The limitations of LEACH motivated researchers to revisit and improve the LEACH protocol to adopt QoS requirements

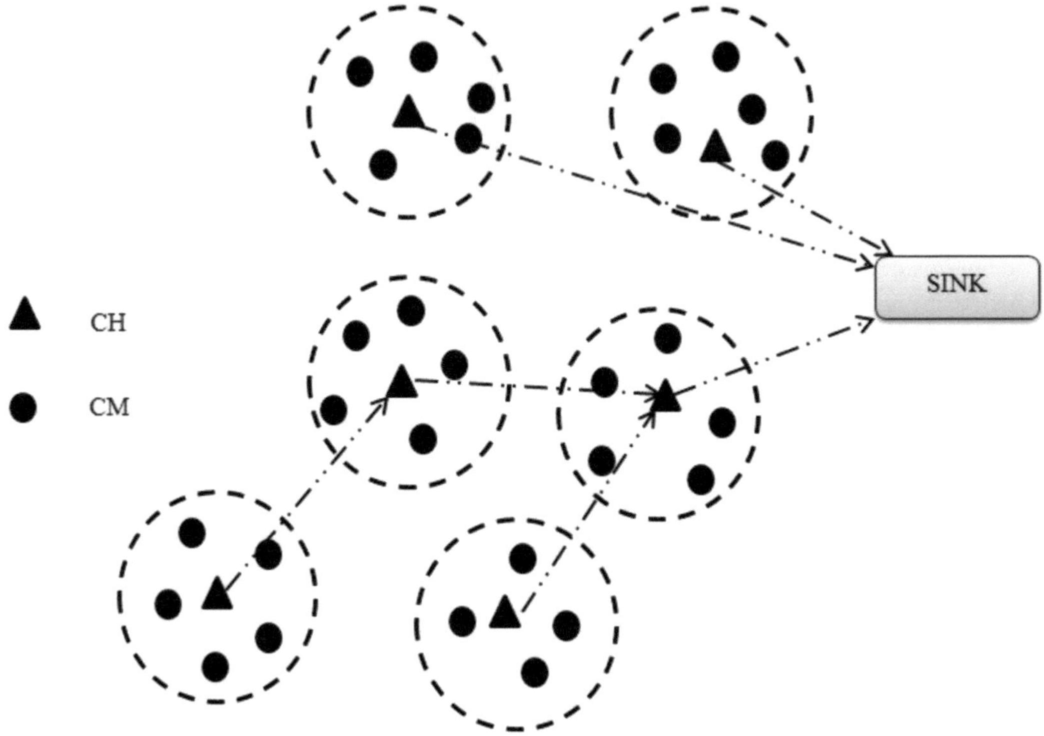

Figure 1. A clustered wireless sensor network.

of WSNs. In general, LEACH and its variants suffer from scalability and load balancing despite their simplicity.

An energy aware clustering protocol using fuzzy logic (ECPF) [3] which is a hierarchical clustering protocol employs a fuzzy based system with the input variables namely, the node degree and the node centrality to form on-demand clusters. This work inspired many researchers to re-estimate the role of cluster heads from the intra-cluster perspective in a hierarchical sensor environment.

A clustering approach namely, energy aware clustering scheme with transmission power control for sensor networks (EACLE) [4] presents a distributed approach for path selection to reach the sink. This scheme sets different levels of transmission power for intra-cluster and inter-cluster communication to improve energy savings.

An energy harvesting protocol namely, energy harvesting and information transmission protocol (EHITP) [5] estimates the energy to be harvested based on the outage probability.

The aforementioned contemporary clustering approaches for wireless sensor networks indicate the need for the new clustering solutions from multiple perspectives.

The advent of new technologies and emerging trends in application development challenge the research findings of performance in WSNs, especially from the energy perspective. A family of solutions is needed to work with various types of unattended sensor networks where many parameters are unpredictable due to the random deployment of sensor nodes. Observing closely, focus is required more on the communication overheads of sensor nodes. Also, any proposed clustering approach should be tested for its scalability in a WSN environment.

1.3. Organization of the chapter

The remaining part of this chapter presents the four proposed clustering approaches for wireless sensor networks. These approaches have improved the lifetime of sensor networks by taking advantages of techniques such as cognitive cluster head selection, support for energy harvesting, hierarchical clustering and effective sleep scheduling of sensor nodes. These have been presented in the following sections, titled, energy aware fuzzy clustering algorithm (EAFCA) [6], efficient energy harvesting assisted clustering (EEHC) algorithm [7], energy-efficient recursive clustering (EERC) algorithm [8] and adaptive distributed clustering algorithm (ADCA) [9], respectively.

2. Energy aware fuzzy clustering algorithm (EAFCA)

2.1. Effective cluster head election in EAFCA

Energy aware fuzzy clustering algorithm (EAFCA) is a proposal based on cognitive technique for non-probabilistic clustering process. In this, the sensor nodes are assumed to be deployed in an unmanned wireless sensor application and clustered from the energy perspective.

The following assumptions have been made on the experimental environment.

- The sensor deployment is done in a random manner.
- It is an unmanned sensor environment.
- All the sensor nodes are kept static.
- The distance between two sensor nodes is measured through the received signal strength.

2.2. Cluster formation

After the sensor nodes are deployed, the distance between any two sensor nodes have to be computed. For this, our proposed approach employs a mechanism in which a beacon signal is transmitted from the sink to the rest of the sensor nodes. Based on the received signal strength, a sensor node can calculate its distance from the sink. Then a group of tentative cluster heads (TCHs) are elected from the sensor network for a specific fraction of the entire network as follows. A threshold "T" is calculated and forwarded to all the sensor nodes. Every sensor node generates a random number and compares the same against the received threshold value. Suppose the generated value is more than the threshold, then the sensor node declares itself as a cluster head. Otherwise, it becomes an ordinary sensor node.

The proposed method results in 2-hop cluster formation and a permanent cluster head (CH) is elected based on fuzzy logic which emphasizes the following three factors:

(1) Remaining residual energy:

This parameter is expected to be higher for an eligible CH in a competition phase as it is heavily engaged to intra-cluster and inter-cluster data traffic.

(2) Node degree at its 2-hop coverage:

This parameter signifies the total number of neighbors which are located in the 2-hop distance from the tentative CH and is calculated as in Eq. (1). It is desirable for a tentative cluster head to have a higher value for this parameter to become a permanent cluster head.

$$2 - hop\, node\, degree = \frac{|s_{2-hop-nbr(i)}|}{\#\, nodes} \tag{1}$$

where S_2-hop-nbr(i) gives the total number of neighbors for the tentative cluster head in its 2-hop coverage. A typical 2-hop clustering environment in wireless sensor networks is represented in **Figure 2** [6].

(3) Centrality of the CH:

For an effective CH, this parameter should yield low values to reduce energy consumption during the data aggregation and flooding processes. Centrality of a node is calculated using Eq. (2).

$$Node\, Centrality = \frac{\sqrt{\frac{\left(\sum_{j \in s_{2-hop-nbr(i)}} d^2(i,j)\right)}{|s_{2-hop-nbr(i)}|}}}{A} \tag{2}$$

where, the parameter "d(i,j)" represents the distance between nodes "i" and "j" in which node j is a member of the set 2-hop-nbr. The variable "A" represents area of the network.

Using tentative cluster heads (TCHs) that are identified and by employing fuzzy logic, primary cluster heads (PCHs) are elected considering the aforementioned three parameters. The fuzzy output variable "chance" is computed for every tentative cluster head as shown in **Table 1** to compute its probability to become to a permanent cluster head.

Using Mamdani method, the fuzzy if-then rules are processed. Every tentative cluster head obtains its chance values and broadcasts the same to all its 2-hop neighbors. It happens in subsequent stages of hop coverage.

Thus a sensor node can either become a cluster head or remains as an ordinary node. An ordinary node has to identify the suitable cluster to join. Suppose it has received advertisements from two or more cluster heads to join, then it will choose the nearby cluster head and joins its group. This is done from the energy perspective. On certain occasions, it may receive advertisements from two cluster heads which are located at equal distances. In such case, the sensor node will choose the cluster head from whom the advertisement has been received earlier and joins. Thus overlapping of clustering is avoided.

Following the process of cluster formation, the data is generated from the sensor nodes. The cluster heads collect the data from the cluster nodes and aggregate the data. Here, a cluster is

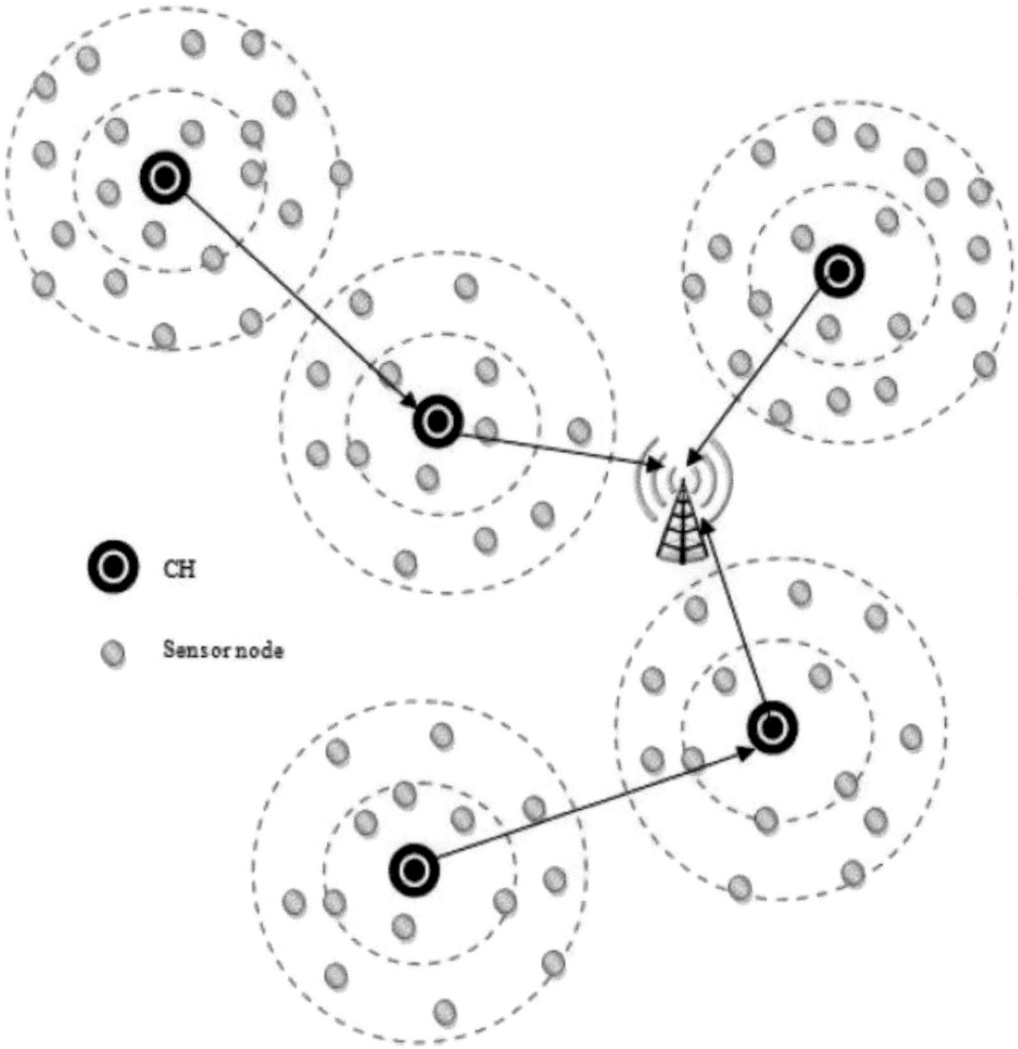

Figure 2. A 2-hop clustered wireless sensor network.

formed on 2-hop coverage and hence the aggregation can be done on a larger scale compared to the conventional 1-hop clustering approaches.

The cluster heads have to report the aggregated data to the sink. Unlike LEACH and many conventional algorithms, the proposed algorithm inherits the presence of multi-hop relay between the cluster head and the sink. This multi-hop relay favors the selection of any one path based on certain probability among multiple choices and the selected path is not necessarily to be an optimal path as followed in many approaches. The repetitive employment of a few popular paths which have been identified as efficient paths causes energy depletion of nodes on these paths and pushes them die soon. Our idea of selecting less used or unused

Energy	2-hop ND	Node centrality	Chance	Energy
Low	Low	Far	Very weak	Low
Low	Low	Medium	Weak	Low
Low	Low	Close	Little weak	Low
Low	Medium	Far	Weak	Low
Low	Medium	Medium	Little weak	Low
Low	Medium	Close	Little weak	Low
Low	High	Far	Little weak	Low
Low	High	Medium	Little weak	Low
Low	High	Close	Medium	Low
Medium	Low	Far	Little weak	Medium
Medium	Low	Medium	Little medium	Medium
Medium	Low	Close	Medium	Medium
Medium	Medium	Far	Little medium	Medium
Medium	Medium	Medium	Medium	Medium
Medium	Medium	Close	High medium	Medium
Medium	High	Far	Medium	Medium
Medium	High	Medium	High medium	Medium
Medium	High	Close	Little strong	Medium
High	Low	Far	Medium	High
High	Low	Medium	High medium	High
High	Low	Close	Little strong	High
High	Medium	Far	High medium	High
High	Medium	Medium	Little strong	High
High	Medium	Close	Strong	High
High	High	Far	Little strong	High
High	High	Medium	Strong	High
High	High	Close	Very strong	High

Table 1. Fuzzy sets for input and output variables.

paths can effectively contribute to the distribution of energy consumption and prolong the lifetime of sensor nodes.

2.3. Performance evaluation of EAFCA

The performance of the algorithm is evaluated under three different scenarios. In scenario 1, the sink is positioned at the center and 100 sensor nodes are deployed. In scenario 2, the sink is positioned at the center while the number of sensor nodes deployed was doubled as 200 to test the scalability of the network. In scenario 3, the sink is located outside the predefined WSN boundaries and the network size is maintained as in scenario 2. The performance of EAFCA algorithm is compared against the benchmarking protocols namely, low energy adaptive clustering hierarchy (LEACH), energy aware clustering protocol using fuzzy logic (ECPF) and energy aware clustering scheme with transmission power control for sensor networks (EACLE).

The metrics employed for computing the lifetime of sensor networks are, first node dies (FND) and half of the nodes alive (HNA) metrics. These two metrics are widely adopted based on the viewpoint whether the energy depletion of the very first node in the network or half of the nodes indicate the death of the network. From the simulation results, it has been observed that the proposed algorithm shows 88% energy improvement compared to LEACH, 46% of improvement with respect to EACLE and 30% of improvement with respect to ECPF.

It is to be observed that LEACH shows the poorest performance among the selected clustering approaches since it does not re-elect cluster heads from the energy perspective and it continues to be a probabilistic model. EACLE shows some improvement in energy consumption as studied from the simulation results. The gain in energy efficiency is achieved in EACLE since it employs multiple paths for inter-cluster traffic and postpones the death of sensor nodes. ECPF claims more energy improvement since it adopts a fuzzy based cluster head election. The results observed across both FND and HNA metric confirms this claim.

The proposed EAFCA reduces energy efficiency by considering necessary and sufficient parameters for a cluster head election and assumes a feasible configuration in which 2-hop coverage is given for every cluster head and multi-hop relay is done for inter-cluster communication. The results demonstrate that EAFCA keeps WSN functioning for longer time than the other approaches.

This work stands as a representative for cognitive and effective cluster head election process. Such strategies expose the sensor networks and its applications to the emerging era of explorations and can be eventually commercialized.

3. Efficient energy harvesting assisted clustering (EEHC) algorithm

3.1. Energy harvesting in wireless sensor networks

Lifetime of wireless sensor applications depends upon the lifetime of the sensor nodes which are constrained by their energy resources. This can be managed by the use of energy harvesting, utilizing ambient sources to prolong the life of the batteries in wireless sensor nodes. The efficiency of this approach depends upon how much energy is harvested. This can majorly influence the lifetime of sensor nodes and in turn that of the sensor network. In our efficient energy harvesting assisted clustering (EEHC) for wireless sensor networks, the effective energy harvesting for wireless sensor networks is experimented and studied through an efficient energy budgeting. The measurement of energy consumption, energy budgeting and energy harvesting are presented as follows.

3.2. Energy consumption

A sensor node consumes energy during sensing the data and forwarding it to the cluster head. A cluster head consumes the energy during the reception, data aggregation and forwarding the aggregated data. The energy consumption of a cluster member to sense and transmit 1-bit of information to the cluster head is estimated in Eq. (3):

$$E_{SN} = S_{ensing} + E_{tx} \tag{3}$$

Assuming a sensing rate of "x," the total data sensed and transmitted by "n" cluster members in a time period "t" is estimated as given in Eq. (4).

$$E_{CM} = (n.x.t).E_{SN} \tag{4}$$

Since the maximum number of cluster members are located at 1-hop distance to the cluster head, it is assumed that the data sensed at time "t" is transmitted to the cluster head within the same interval. Suppose a cluster head collects L-bit length of data at time "t," (i.e., L= x.t.n) then the total energy conservation for data reception, aggregation and forwarding in that CH across time period "t" is estimated as given in Eq. (5).

$$E_{CH} = n.x.t.E_{rx} + n.x.t.E_{DA} + \frac{n.x.t}{\alpha}.E_{txr} \tag{5}$$

where α stands for aggregation ratio.

The total energy consumed in time "t" for a cluster is given in Eq. (6).

$$E_c = E_{CH} + E_{CM} \tag{6}$$

For a time slot "t," the entire cluster, i.e., all the cluster nodes including the cluster head should harvest the energy equal to that of the estimated energy. Suppose there are "n" cluster members and a cluster head, then the energy that is required to be harvested by a sensor node in a cluster is given by the Eq. (7).

$$E_h = \frac{E_c}{n+1} \tag{7}$$

3.3. Energy harvestings

For every time interval "T" between time "t_1" and "t_1+T," the harvested energy is calculated as given in Eq. (8).

$$E(t_1 \text{ to } t_1 + T) = E_t(n) + \tau\int_{t_1}^{t_1+T} E_h(n,t)\,dt - \int_{t_1}^{t_1+T} E_l(n,t)\,dt \tag{8}$$

In Eq. (8), the three components represent the energy of the node at starting time "t_1," energy harvested at time interval "T" and energy leakage during this interval. The factor "τ" represents charging efficiency. All the sensor nodes are provided with the storage buffers to store the harvested energy.

3.4. Energy budget

The energy consumed must be compensated by the energy harvested within the boundary of a cluster in any given time slot, i.e., the energy budget should harvest more energy than that of the energy consumed in every cluster periodically. An efficient energy budget should ensure that the energy consumption should not be increased than the energy harvested across any time slice.

3.5. Performance evaluation of EEHC

The performance of our EEHC has been compared with a modern clustering protocol named energy harvesting and information transmission protocol (EHITP) and the classical clustering protocol LEACH under three scenarios. In scenario 1, 100 sensor nodes are deployed in a region of 200×200 m^2. In scenarios 2 and 3, the population of sensor nodes and area are doubled successively. The experimental results indicate that EEHC exhibits a mean improvement of 91 and 67% when compared to LEACH and EHITP, respectively.

Thus the harvesting can be made efficient through appropriate budgeting in wireless sensor networks and this budgeting further is influenced by the nature of the sensor applications and critical dynamic constraints.

4. Energy-efficient recursive clustering (EERC) algorithm

4.1. Event based clustering in wireless sensor networks

Generally, wireless sensor networks are employed for two purposes: continuous data monitoring and event monitoring. An example for the former category is a weather monitoring sensor network that measures temperature, moisture, etc. A typical event-based sensor network is habitat monitoring such as surveillance of wild animals and smart home applications. Except a few applications in the second category, most of the senor nodes are put on sleep mode in order to save power and the effective sleep/wakeup scheduling algorithms are required to determine the set of sensor nodes that can be scheduled to sleep with respect to time.

The energy-efficient recursive clustering (EERC) algorithm is an event-driven clustering algorithm, i.e., on the occurrence of an event, the clusters are formed to reduce energy dissemination, in a recursive fashion. The recursive clustering approach employs two stages of clustering process. The first stage of clustering is followed by further partitioning of clusters. Then CHs are elected from energy perspective.

4.2. Recursive clustering approach

The recursive clustering approach employs two stages of clustering process. After the deployment of sensor nodes, the distance between the nodes is computed using Euclidean distance. Based on the distance, "k" number of clusters are formed which results in the first stage of clusters. Since the clustering process is modeled as recursive process further the "k" number of clusters is divided in "j" number of clusters based on the distance and interval between the nodes which leads to the second stage of clusters. A typical process of this two-stage clustering is pictorially represented as shown in **Figure 3**. After recursive clustering process, CH is elected based on energy levels and employing round robin scheduling algorithm. Based on the computations, the node with minimum turnaround time and high energy is elected as the CH among the nodes in the cluster.

Each node senses the data for every two rounds. After the completion of two rounds, turnaround time is calculated. The node having the minimum turnaround time is elected as the cluster head among the nodes in the cluster. The sensed data from each node is sent to the cluster head. In the cluster head, data is aggregated and sent to the base station by the multi-hop routing. The aggregated data in a cluster head leads to less transmission data, decrease in overheads and decrease in energy consumption.

4.3. Performance evaluation of EERC

The performance of our EERC has been evaluated against the conventional LEACH clustering approach under three scenarios through simulation. In scenario 1, the number of sensor node deployed is 100 and in order to test the scalability the scenario 2 and scenario 3 were considered. In scenario 2, 250 sensor nodes and in scenario 3, 500 sensor nodes have been considered.

From the results obtained for scenario 1, our EERC approach shows performance improvement when compared to the classic LEACH protocol. For 100 nodes, there is 23.85% increase

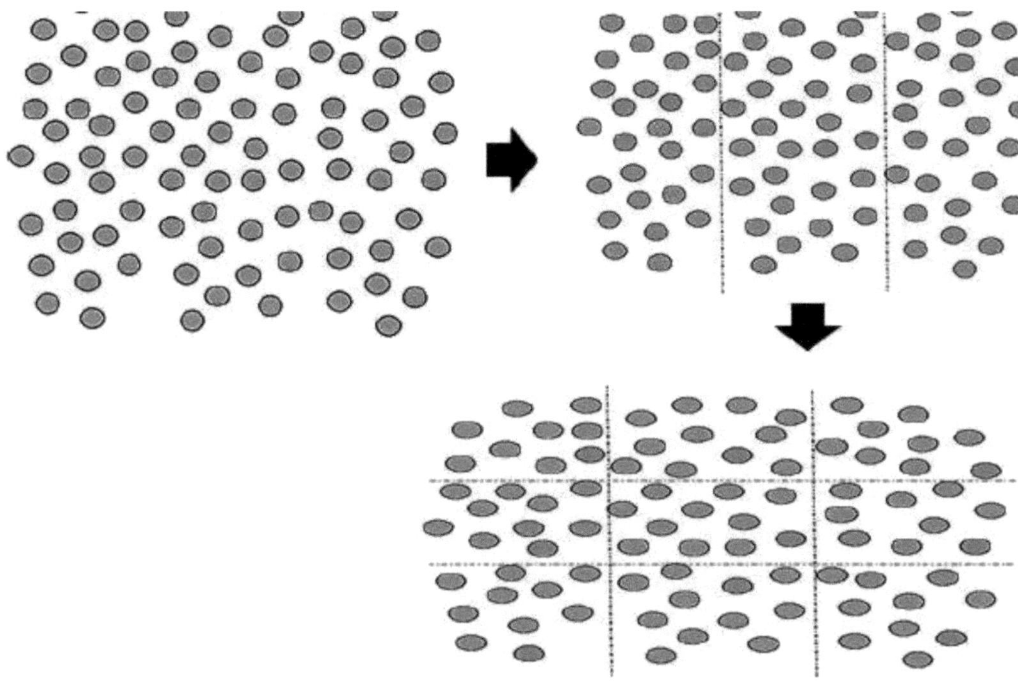

Figure 3. Recursive clustering in wireless sensor networks.

in lifetime, 0.287% decrease in energy consumption, 2.522% decrease in delay, 1.497% increase in transmission time, 1.8% increase in goodput and 0.56% decrease in overhead. In scenario 2, EERC shows performance improvement of 12.58% increase in lifetime, 0.402% decrease in energy consumption, 9.815% decrease in delay, 9.289% increase in transmission time, 0.524% increase in goodput and 0.554% decrease in overhead. In scenario 3, EERC exhibits performance improvement of 11.03% increase in lifetime, 0.619% decrease in energy consumption, 5.735% decrease in delay, 9.289% increase in transmission time, 0.524% increase in goodput and 0.554% decrease in overhead throughput.

Thus the recursive clustering technique gives considerable improvement across various performance factors, sustaining the equilibrium of the entire network.

5. Adaptive distributed clustering algorithm (ADCA)

5.1. Similarity measure in sensor data

Similarity Measure is the metric that is employed in this approach for clustering the sensor nodes from a temporal and spatial perspective. The sensor nodes of the same neighborhood produce similar data. Similarly, the data generated from the same sensor node may exhibit similarity to considerable extend except exceptional scenarios of an application. Also, the data that is generated from the same sensor node on successive timeslots in the period of observation may reveal similarity. The redundancy of the data generation and aggregation is effectively controlled through this technique which considerably contributes to the energy consumption in sensor networks.

The component of similarity measure amidst the data sensed from sensor nodes opens the door to devise an effective sleep schedule and save energy. This idea is capitalized in the proposed adaptive distributed clustering algorithm (ADCA).

5.2. Phases in adaptive distributed clustering algorithm

It employs two major phases: a cluster formation phase and an adaptive sleep duty cycle phase. In the cluster formation phase, the data generation rate and the similarity between data series are analyzed by the sink. Based on estimation, the nodes are grouped into various clusters. In each cluster, the cluster heads are selected based on the connectivity and residual energy.

In practical scenarios, the clusters may not be in equal size. Based on the similarity measure of the time series, the clustering is done in this approach. By using the similarity measure, the degree of spatial correlation can be calculated. Generally, for two locations with high spatial correlation, their corresponding time series are associated with a high similarity measure. Hence, in a very smooth sub-region, the observed measure has only small changes within the sub-region. In other words, the difference between the observations at any two locations within the sub-region may be very small and hence negligible.

Therefore, without compromising the observation reliability, the working sensor nodes within this sub-region could be sparse. On the other hand, the working sensor nodes should be dense in a fast changing sub-region. The spatial sampling rate has to match the spatial variation of the observed physical incident by setting an appropriate similarity measure threshold value. Hence, the similarity measure threshold value includes a degree of independency. This value can be tuned to balance the trade-off between reliability and energy consumption.

In the sleep duty cycle phase, the data generation rates of cluster members are compared with a minimum threshold level. The nodes which have rates lower than the threshold level are cumulatively allotted a sleep duty cycle for a predefined period. The sleep/wakeup schedule is informed to every sensor node. The scheduling is done on a fair and distributed manner to regulate energy consumption.

Every cluster head collects the data from its members and checks for the similarity in the received data. If it encounters a significant level of change, it reports to the sink along with the data. The sink then performs reclustering or rescheduling of sleep duty cycle, if necessary. Thus, a part of the workload of the sink in periodical checking of the nodes is shared by the cluster heads.

The data values of two sensor nodes n_i and n_j are said to be similar, if

$$mt_i = mt_j \tag{9}$$

where mt_i and mt_j are the magnitudes of the values of n_i and n_j

$$D_{ij} < D_{Th} \tag{10}$$

where D_{ij} is the distance between n_i and n_j and D_{Th} is the distance threshold.

$$\delta_{ij} < \delta_{min}, \delta_R = ABS(R_j - R_i) \tag{11}$$

where R_i and R_j are the sending rates of n_i and n_j, respectively.

Sending rates are calculated in Eqs. (12) and (13).

$$R_i = NP_i/T \tag{12}$$

$$R_j = NP_j/T \tag{13}$$

here, NP_i is the number of packets sent by sensor node i in a time period T.

δ_{ij} is the absolute difference of sending rates of n_i and n_j

δ_{min} is the minimum threshold value for δ_R.

The two nodes can be represented as points in three dimensions. Node i has a set of coordinates (mt_i, D_i and R_i) and Node j has coordinates (mt_j, D_j and R_j).Therefore the Similarity Distance between the nodes n_i and n_j is given by the Eq. (14).

$$SM_{ij} = \sqrt{\sum_{k=1}^{n} (x_{ik} - x_{jk})^2}, \ n = 3. \quad (14)$$

where, $x_{i1} = mt_i$, $x_{i2} = D_i$, $x_{i3} = R_i$ and $x_{j1} = mt_j$, $x_{j2} = D_j$, $x_{j3} = R_j$.

Here, "n" refers to the number of similarity metrics.

5.3. Cluster formation

In our proposed algorithm, the clustering problem is represented as a clique-covering problem. A graph G is created such that each sensor node is a vertex in the graph. An edge (u,v) between nodes u and v is drawn if SM (u,v) > SM_{Th}.

A cluster is observed as a clique in this problem. A greedy algorithm is used to heuristically find the cliques. Until all vertices are covered, the search starts from the vertex with the largest node degree. The output of this algorithm is a set of cliques that cover all vertices.

5.4. Performance evaluation of ADCA

The performance of the proposed ADCA algorithm has been compared against a contemporary clustering algorithm named energy-efficient distributed clustering (EEDC) algorithm [10]. Number of sensor nodes has been varied from 20 to 100 in a simulation area of 500×500 m^2. The results obtained show the performance improvement gained by ADCA with respect to EEDC in terms of energy consumption (25%), delay (18%) and delivery ratio (20%) from the mean measurements of multiple runs.

Thus the effective sleep duty cycle considerably reduces the energy consumption in wireless sensor networks ensuring that the overhead and delay are not increased under such scenarios.

6. Summary

Wireless Sensor networks are more sophisticated in their requirements and to provide clustering solutions for them requires adequate knowledge on the nature of the applications, capacity limitations of sensor nodes, the tradeoff among the expected performance parameters and the limitations of emerging technologies.

This chapter has outlined four modern clustering approaches (EAFCA, EEHC, EERC and ADCA) designed for wireless sensor networks, each from a distinguishable perspective. The simulation results obtained for the proposed clustering approaches are encouraging. This will kindle researchers to explore further in this area. The journey of research in this field has crossed significant milestones, yet it has been left with many open-ended issues and unexplored roads due to the presence of inherent trade-offs among the performance factors and dynamic needs of sensor applications. The performance of a wireless sensor network can further be explored through holistic approaches invented or inherited from modern technological advancements.

Author details

I.S. Akila[1]*, S.V. Manisekaran[2] and R. Venkatesan[3]

*Address all correspondence to: akila_subramaniam@yahoo.co.in

1 ECE Department, Coimbatore Institute of Technology, Coimbatore, India

2 IT Department, Anna University Regional Campus, Coimbatore, India

3 CSE Department, PSG College of Technology, Coimbatore, India

References

[1] Abbasi AA, Younis M. A survey on clustering algorithms for wireless sensor networks. Computer Communications. 2007;30(14–15):2826-2841

[2] Heinzelman WB, Chandrakasan AP, Balakrishnan H. Energy-efficient communication protocol for wireless microsensor networks. Proceedings of the International Conference on System Sciences (HICSS); Wailea Maui, Hawaii; 2000. pp. 10-19

[3] Taheri H, Neamatollahi P, Younis OM, Naghibzadeh S, Yaghmaee MH. An energy-aware distributed clustering protocol in wireless sensor networks using fuzzy logic. Ad Hoc Networks. 2012;10(7):1469-1481

[4] Yanagihara K, Taketsugu J, Fukui K, Fukunaga S, Hara S, Kitayama K. EACLE: Energy-aware clustering scheme with transmission power control for sensor networks. Wireless Personal Communications. 2007;40(3):401-415. DOI: 10.1007/s11277-006-9199-2

[5] Zhang XF, Yin C. Energy harvesting and information transmission protocol in sensors networks. Journal of Sensors. 2016;2016:1-5. DOI: 10.1155/2016/9364716

[6] Akila IS, Venkatesan R. A cognitive multi-hop clustering approach for wireless sensor networks. Wireless Personal Communications. 2016;90(2):729-747. DOI: 10.1007/s11277-016-3200-5

[7] Akila IS, Venkatesan R. An efficient energy harvesting assisted clustering scheme for wireless sensor networks. International Journal on Recent and Innovation Trends in Computing and Communication. 2016;4(4):548-558

[8] Akila IS, Subaselvi S. Energy efficient recursive clustering approach for wireless sensor networks. International Journal of Electronics, Electrical and Computational System. 2017;6(5):121-130

[9] Manisekaran SV, Venkatesan R. An adaptive distributed power efficient clustering algorithm for wireless sensor networks. . American Journal of Scientific Research. 2010;20:50-63

[10] Liu C, Wu K, Pei J. An energy-efficient data collection framework for wireless sensor networks by exploiting spatiotemporal correlation. IEEE Transaction on Parallel and Distributed Systems. 2007;18(7):1010-1023. DOI: 10.1109/TPDS.2007.1046

Chapter 9

Fuzzy Adaptive Setpoint Weighting Controller for WirelessHART Networked Control Systems

Sabo Miya Hassan, Rosdiazli Ibrahim, Nordin Saad, Vijanth Sagayan Asirvadam, Kishore Bingi and Tran Duc Chung

Additional information is available at the end of the chapter

http://dx.doi.org/10.5772/intechopen.70179

Abstract

Gain range limitation of conventional proportional-integral-derivative (PID) controllers has made them unsuitable for application in a delayed environment. These controllers are also not suitable for use in a Wireless Highway Addressable Remote Transducer (WirelessHART) protocol networked control setup. This is due to stochastic network-induced delay and uncertainties such as packet dropout. The use of setpoint weighting strategy has been proposed to improve the performance of the PID in such environments. However, the stochastic delay still makes it difficult to achieve optimal performance. This chapter proposes an adaptation to the setpoint weighting technique. The proposed approach will be used to adapt the setpoint weighting structure to variation in WirelessHART network-induced delay through fuzzy inference. Result comparison of the proposed approach with both setpoint weighting and proportional-integral (PI) control strategy shows improved setpoint tracking and load regulation. For the first-, second- and third-order systems considered, analysis of the results in the time domain shows that in terms of overshoot, undershoot, rise time, and settling times, the proposed approach outperforms both the setpoint weighting and the PI controller. The approach also shows faster recovery from disturbance effect.

Keywords: setpoint weighting, fuzzy adaptation, WirelessHART, PID, wireless sensor networks

1. Introduction

Recent advances in wireless technology have prompted researchers to look into its application for industrial process monitoring and control. However, this attempt was hindered by lack of an open and interoperable industrial standard [1–4]. This changed with the coming on board of standards such as WirelessHART, Wireless Networks for Industrial Automation-Process Automation (WIA-PA) and International Society of Automation (ISA) wireless (ISA100.11a). Of these three standards, the WirelessHART has upper hand since it is based on the well-known Highway Addressable Remote Transducer (HART) protocol that is already established with millions of HART-enabled devices already installed worldwide [5–7]. The WirelessHART standard protocol is based on the Open Systems Interconnection model (OSI model) as shown in **Figure 1**.

The WirelessHART standard adopted a modified version of the physical layer of the IEEE 802.15.4-2006 and operates on the 2.4-GHz industrial, scientific and medical (ISM) radio frequency band. The signals are transmitted over this frequency using 15 channels spaced 5 MHz apart. The time division multiple access (TDMA) method is used for communication whereby packets are sent using 10 ms time slots arranged in the form of superframe. Each superframe thus consists of trains of 10 ms time slots (**Figure 2**). To avoid interference of other networks and multi-path fading, the standard adopts the strategy of channel hopping between its 15 channels [5, 8]. The standard is secured using the industry standard AES-128 ciphers and keys. The mesh topology of the standard makes it highly reliable, self-organizing and self-healing. In addition to the host computer, a typical WirelessHART network consists of at least a gateway, network manager and field devices as shown in **Figure 3**.

In spite of the advantages of reduced cabling, improved reliability, scalability and many more offered by wireless technology such as WirelessHART, its application for control is still faced with the challenges of network-induced stochastic delays and uncertainties such as packet

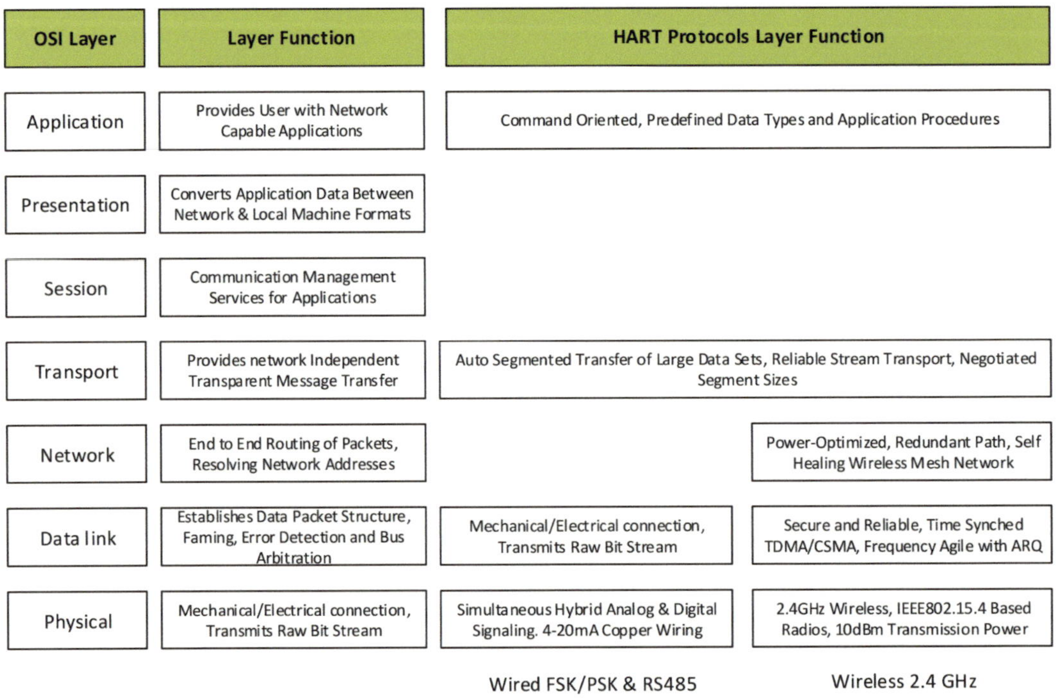

Figure 1. WirelessHART protocol based on OSI layers.

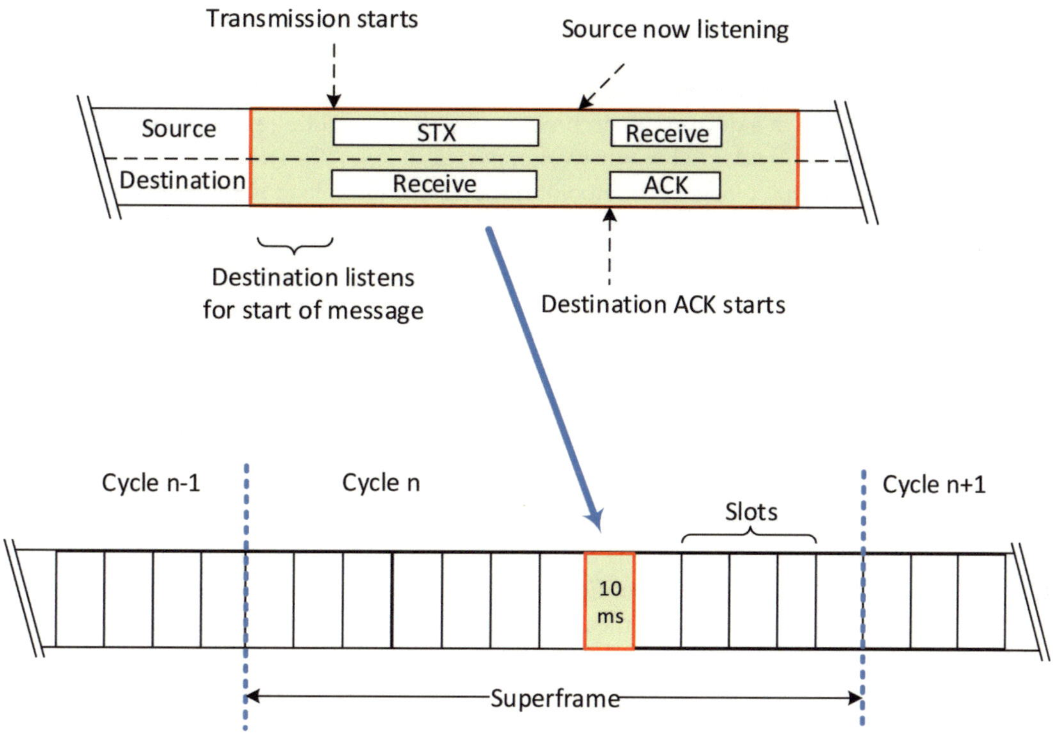

Figure 2. WirelessHART superframe structure.

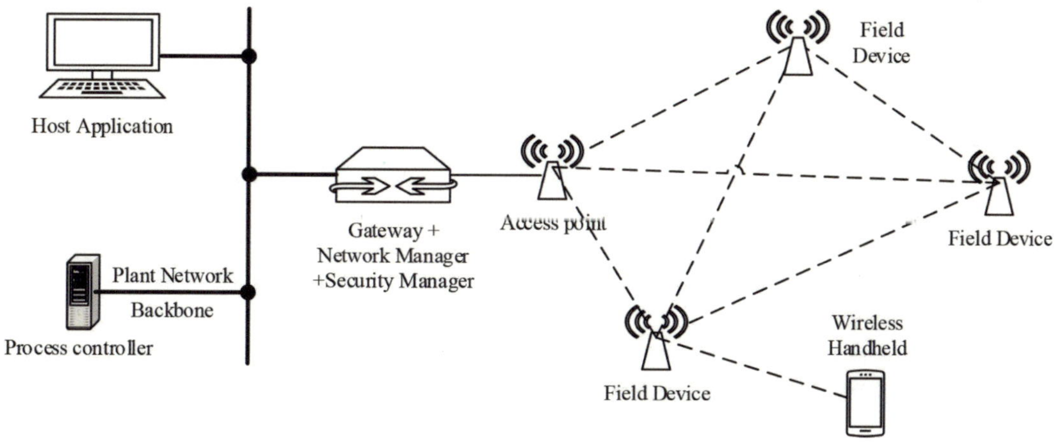

Figure 3. Typical WirelessHART network.

dropout. This is as a result of the use of wireless transmitters in the network, which transmit signals aperiodically [9, 10].

From the control perfective, the most common controllers used in the industry are the PID controllers. These controllers are, however, inadequate to be used in a delayed environment [11].

This is because long delays cause oscillation in the response of the system controlled with PID. Furthermore, the PID is limited in gain range, which makes it difficult to adapt to the stochastic nature of the delays in the WirelessHART environment [12]. In an attempt to improve on the performance of the PID in a delayed environment, a setpoint weighting structure was proposed in Ref. [11]. This was later adopted in our work reported in Ref. [13]. The design allows for two degree of freedom control, where both setpoint tracking and good load regulation are achieved. However, if the variability of the network delay is high or if the plant to be controlled is of higher order, the setpoint weighting strategy fails to give optimal performance. Thus, this chapter proposes the adaptation of the setpoint weighting control strategy to the stochastic delay through fuzzy inference system. Fuzzy gain tuning has been an effective way to tune parameters of a controller online with respect to parameter changes. It has been applied recently to tune PID controller for multiple input multiple output (MIMO) systems [14], continuous stirred-tank reactor (CSTR) systems [15], maximum power point tracking in a photovoltaic system [16], load frequency control [17, 18] and many other control applications [19–22].

Among the key advantages of the proposed approach is that although the model of the process to be controlled may be required for the design, it is however not mandatory. Furthermore, in the design, original PID feedback configuration is retained; thus, no modification of the existing structure is required. Finally, the gain range of the PID is significantly extended while achieving robust performance even with external disturbances.

The reminder of this chapter is organized as follows: in section 2, the methodology for the delay measurement is presented, while section 3 gives the design of fuzzy adaptation scheme. The results are presented and discussed in section 4, while in section 5 conclusion is drawn.

2. WirelessHART network delay measurement

WirelessHART network delay is measured using Dust Networks DC9007A SmartMesh starter kits produced by Linear Technology. The experimental schematic is shown in **Figure 4**. The experimental setup consists of a host computer, LTP5903CEN-WHR WirelessHART network manager/Gateway and DC9003-C Eterna WirelessHART motes. As seen from the schematic, the host computer is connected to the gateway through RJ-45 cable, while communication between the gateway and the motes is achieved wirelessly. In this setup, each mote is assumed to be connected to a process plant. Thus, to measure the upstream delay from gateway to the mote t_u, and the downstream delay from mote to the gateway t_d,

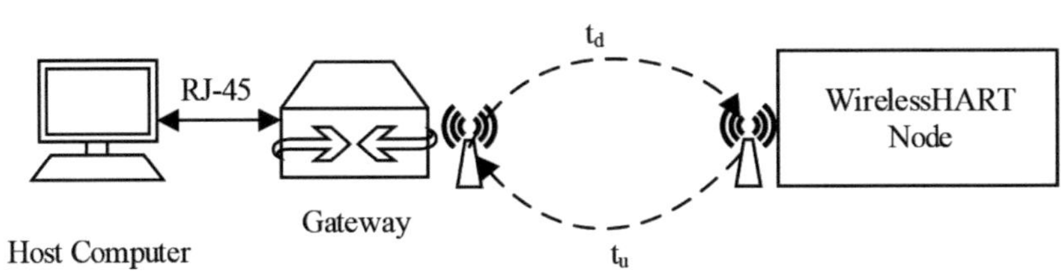

Figure 4. WirelessHART network delay measurement schematic.

two-step procedures are involved. First the delay is obtained in the gateway by executing command *exec getLatency MACaddress* in gateway, where *MACaddress* is the MAC address of the node in the gateway [13]. Secondly, this delay information is obtained in MATLAB from gateway through the use of Secure Shell (SSH2) software. This is achieved by establishing a secured communication between MATLAB in host and the gateway. The SSH2 command used for this purpose is *ssh2_config ('IP address,' 'userName,' 'password')*. The complete procedure is shown in **Figure 5**.

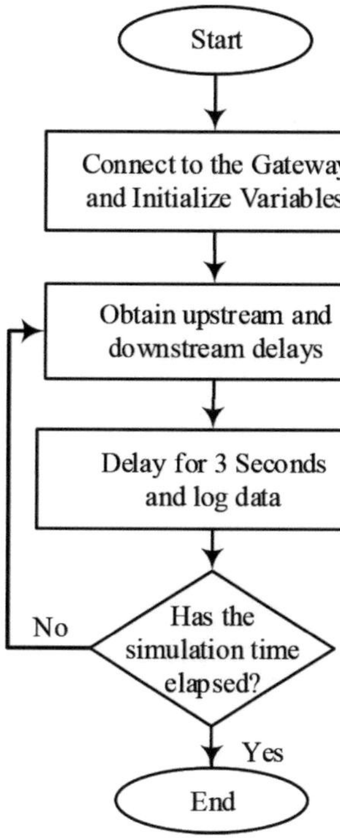

Figure 5. Procedure for delay measurement.

3. Fuzzy adaptive setpoint weighting structure for WirelessHART system (FASW)

This section details the complete design procedure for the fuzzy adaptive setpoint weighting (FASW) control strategy. To do this, the setpoint weighting (SW) structure will first be designed. Then, the fuzzy adaptation will be incorporated to form the FASW structure.

3.1. Setpoint weighting structure

Considering the plant $G(s)$ of Eq. (1) in a WirelessHART environment, the typical setpoint weighting strategy for the system as reported in Ref. [13] is shown in **Figure 6**.

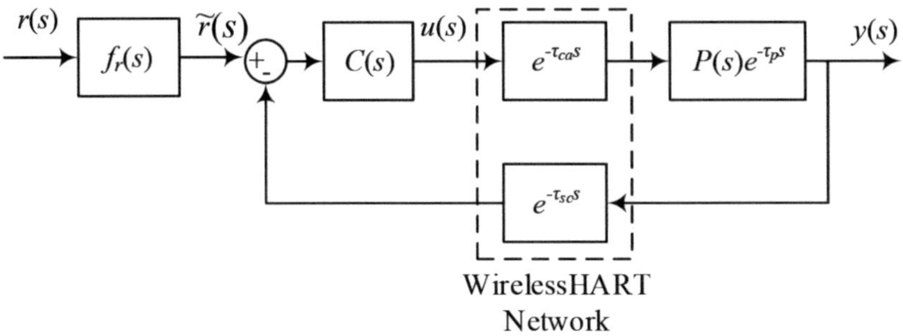

Figure 6. WirelessHART network setpoint weighting structure.

$$G(s) = P(s)e^{-\tau_p s} = \frac{K_p}{1 + sT} e^{-\tau_p s} \tag{1}$$

where K_p, T and τ_p are the plant gain, time constant and dead-time respectively.

From **Figure 6**, the closed-loop transfer function from $y(s)$ to $r(s)$ is given as

$$\frac{y(s)}{r(s)} = \frac{C(s)P(s)e^{-(\tau_{ca}+\tau_p)s}}{1 + C(s)P(s)e^{-(\tau_{ca}+\tau_{sc}+\tau_p)s}} f_r(s) \tag{2}$$

where τ_{ca} and τ_{sc} are controller to actuator delay and sensor to controller delay, respectively. In this work, $\tau_{ca} = t_d$ and $\tau_{sc} = t_u$.

If $\tau_1 = \tau_{ca} + \tau_p$ and $\tau_2 = \tau_{ca} + \tau_{sc} + \tau_p$, then Eq. (2) becomes

$$\frac{y(s)}{r(s)} = \frac{C(s)P(s)e^{-\tau_1 s}}{1 + C(s)P(s)e^{-\tau_2 s}} f_r(s) \tag{3}$$

As reported in our earlier work in Ref. [13], the general setpoint weighting function $f_r(s)$ is given in the following equation

$$f_r(s) = G_r(s) + \tilde{G}_{yr}(s)(e^{-\tilde{\tau} s} - G_r(s)) \tag{4}$$

where $\tilde{G}_{yr}$ is the desired closed-loop response, $G_r(s)$ is the feedforward gain enhancement term, and $\tilde{\tau}$ is the delay estimate. Thus, using Eq. (4) in Eq. (3), we have

$$\frac{y(s)}{r(s)} = \frac{\hat{G}_{yr}(s)e^{-\tau_1 s}(G_r(s) - G_r(s)\tilde{G}_{yr}(s) + \tilde{G}_{yr}(s)e^{-\tilde{\tau} s})}{G_r(s) - G_r(s)\hat{G}_{yr}(s) + \hat{G}_{yr}(s)e^{-\tau_2 s}} \tag{5}$$

where $\hat{G}_{yr}(s) = \frac{G_r(s)C(s)P(s)}{1+G_r(s)C(s)P(s)}$.

Under the conditions $\tilde{\tau} = \tau_2$, $\hat{G}_{yr}(s) = \tilde{G}_{yr}(s)$, and after pole-zero cancellation, Eq. (5) reduces to

$$\frac{y(s)}{r(s)} = \hat{G}_{yr}(s)e^{-\tau_1 s} \tag{6}$$

This indicates that Eq. (6) has decoupled the delay term from the desired closed-loop response $\hat{G}_{yr}(s)$. Thus, the implementation of setpoint weighting function $f_r(s)$ is shown in **Figure 7**.

3.2. Design procedures for SW function

To design the proposed fuzzy adaptation scheme, we will first design the setpoint weighting function as follows:

First, the controller $C(s)$ is a PI controller given by

$$C(s) = K_C\left(1 + \frac{1}{T_i s}\right) \tag{7}$$

where the proportional gain is related to the system parameters as $K_C = \frac{0.5T}{K_p\tau_2}$ and the controller time constant as $T_i = T$.

If $C(s)$ is expressed as $\frac{A_c(s)}{B_c(s)}$, then the feedforward gain enhancement term $G_r(s)$ of $f_r(s)$ is designed as follows

$$G_r(s) = \frac{KC(s)^{-1}P(s)^{-1}}{B_c(s)} \tag{8}$$

where K is a tunable gain.

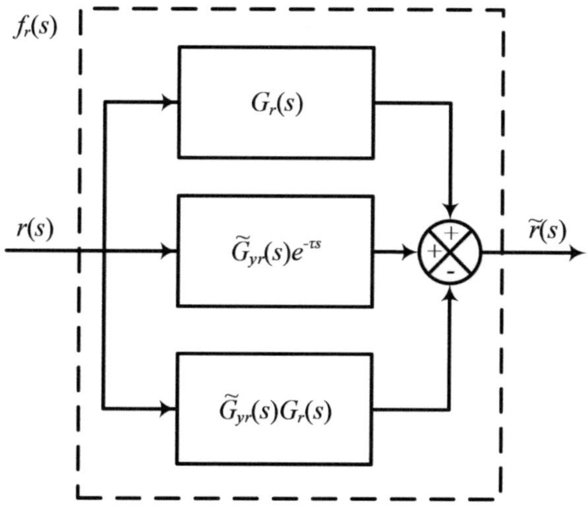

Figure 7. Implementation of setpoint weighting structure.

It should be noted that $G_r(s)$ can be selected simply as K if there is no much information about the system to be controlled.

The desired closed-loop function is thus designed using the following relationship

$$\hat{G}_{yr}(s) = \frac{1}{B_c(s)/K + 1} \tag{9}$$

3.3. Fuzzy adaptation mechanism

If the setpoint weighting function $f_r(s)$ is observed, it can be seen that the terms that depend on the estimate of both the plant dead-time and the network stochastic delay are the gain enhancement term $G_r(s)$ and the delay estimate term $e^{-s\tilde{\tau}}$. Thus, in this work, we will use fuzzy adaption mechanism to adjust these parameters accordingly to ensure smooth setpoint tracking and good load regulation. The proposed adaptation mechanism is shown in **Figure 8**.

The inputs of the supervisor (fuzzy) are the error (e) and its change Δe. The adaptation on $f_r(s)$ is aiming to correct the system evolution while acting on the control law. During on line operation of the controller, the fuzzy system allows for adaptation of the parameters of the SW function. The change in SW parameters ΔK and $\Delta\tau$ is tuned at each sampling time by using fuzzy adaptation as earlier shown in the figure. The respective ranges of the inputs and outputs of fuzzy tuner are as follows:

$$e,\ \Delta e \in [-2, 2]$$

$$\Delta K \in [-2, 2],\ \Delta\tau \in [0, 2]$$

The range is selected based on the information obtained from the variation of the WirelessHART network delay.

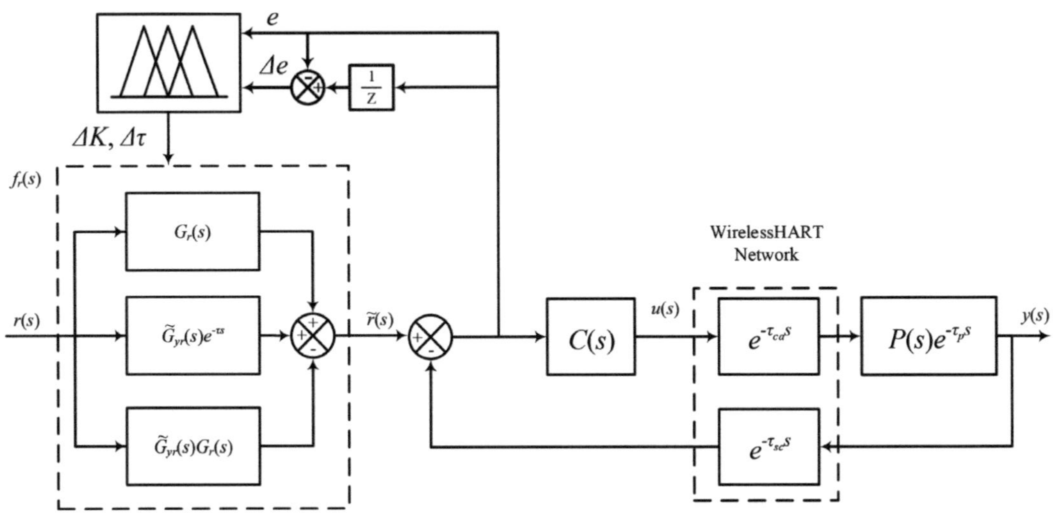

Figure 8. Fuzzy adaptive setpoint weighting structure.

In this proposed fuzzy adaption method, the control rules are developed with the error (e) and change in error (Δe) as a premise and the change in gain (ΔK) and change in delay ($\Delta\tau$) as consequent of each rule. An example of the tuning rule is given as

IF e is NB and Δe is NB, then ΔK is NVB and $\Delta\tau$ is Z.

To achieve smooth adaption, five Gaussian membership functions for input variables and nine Gaussian memberships for output variables have been chosen as shown in **Figure 9**.

The linguistic descriptions of the input membership functions in the figure are Negative Big (NB), Negative Small (NS), Zero (Z), Positive Small (PS), and Positive Big (PB). The output membership functions of ΔK are Negative Very Big (NVB), Negative Big (NB), Negative Medium (NM), Negative Small (NS), Zero (Z), Positive Small (PS), Positive Medium (PM), Positive Big (PB), and Positive Very Big (PVB). Similarly, the linguistic descriptions for the output membership functions of $\Delta\tau$ are Zero (Z), Very Small (VS), Small (S), Small Medium (SM), Medium (M), Small Big (SB), Medium Big (MB), Big (B), and Very Big (VB).

The 25 fuzzy rules are given in **Table 1**. The table is generated based on the rule given above. As seen from the table, the first argument of the output represents ΔK, while the second

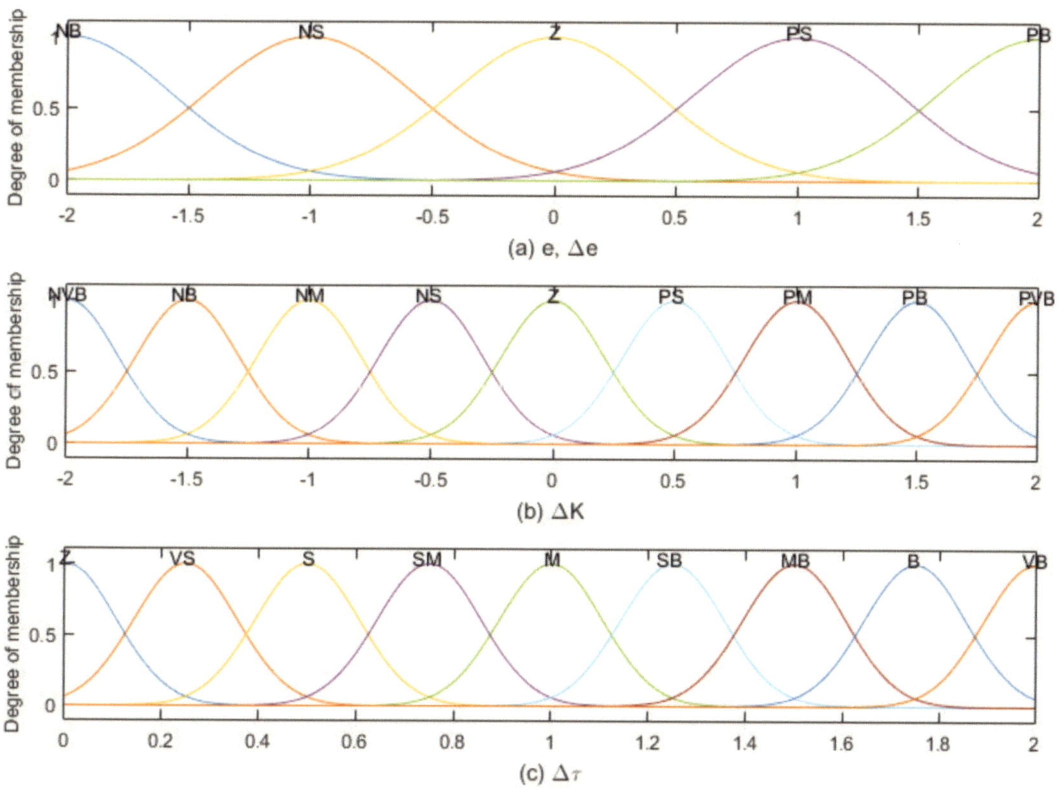

Figure 9. Fuzzy membership functions.

e\Δe	NB	NS	Z	PS	PB
NB	(NVB, Z)	(NB, VS)	(NM, S)	(NS, SM)	(Z, M)
NS	(NB, VS)	(NM, S)	(NS, SM)	(Z, M)	(PS, SB)
Z	(NM, S)	(NS, SM)	(Z, M)	(PS, SB)	(PM, MB)
PS	(NS, SM)	(Z, M)	(PS, SB)	(PM, MB)	(PB, B)
PB	(Z, M)	(PS, SB)	(PM, MB)	(PB, B)	(PVB, VB)

Table 1. Fuzzy rule table.

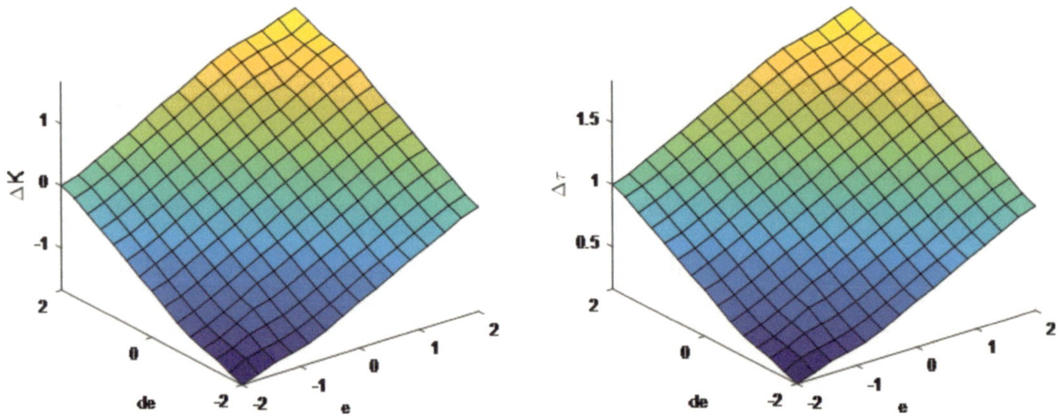

Figure 10. Fuzzy rule surface.

argument represents $\Delta\tau$, i.e., $(\Delta K, \Delta\tau)$. The respective rule surfaces for the two outputs based on **Table 1** are given in **Figure 10**.

Fuzzification is achieved using the intersection minimum operation given as follows

$$\mu_{A\cap B}(x, y) = \min(\mu_A(x, y), \mu_B(x, y)) \tag{10}$$

where A and B are input fuzzy sets (i.e., e and Δe). The values for these inputs are calculated at each sampling time as

$$e(t) = r(t) - y(t) \tag{11}$$

$$\Delta e = \Delta e(t) - \Delta e(t-1) \tag{12}$$

For defuzzification, the commonly used centroid method is selected for finding the crisp value of the output. The centroid method is given as:

$$\mu_o = \frac{\sum_{i=1}^{R} c_i\mu_i}{\sum_{i=1}^{R} \mu_i} \tag{13}$$

http://dx.doi.org/10.5772/intechopen.70179

where

- μ_o is the fuzzy output.
- c_i is the center of the membership function of the consequent ith rule.
- μ_i is the membership value of the premise's ith rule.
- R is the total number of fuzzy rules.

4. Results and discussions

This section will present and discuss the results of the proposed approach. In this chapter, three plant models representing first, second and third orders plus dead-time systems are considered. The transfer functions for these models are given in Eqs. (14), (15) and (16), respectively. The parameters of the various controllers used are shown in **Table 2**. In the table, K_{C1} is the controller gain used for the design of the SW controllers, while K_{C2} is the proportional gain of the PI controller given in Eq. (7). K_{C1} is selected as between 80 and 90% of K_{C2}. The profile and statistical information for the experimental WirelessHART network delay are also given in **Figure 11** and **Table 3**, respectively. Here, the variation in especially upstream delay is observed.

$$P_1 = \frac{1}{1+2s}e^{-4s} \tag{14}$$

$$P_2 = \frac{1}{(s+1)^2}e^{-4s} \tag{15}$$

$$P_3 = \frac{1}{(s+1)^3}e^{-5s} \tag{16}$$

4.1. First-order plant

The setpoint tracking and disturbance rejection response for P_1 with various controller configurations are given in **Figure 12**. From the figure, it can be seen that the setpoint tracking ability and disturbance rejection capability of the two setpoint weighted controllers SW and FASW are

Plant	Parameter				
	$G_r(s)$	$\hat{G}_{yr}(s)$	K_{C1}	K_{C2}	T_i
P_1	13.42	$\frac{1}{2s+1}$	0.1744	0.1938	2
P_2	$\frac{12.05(s^2+2s+1)}{(1.3s+1)(s+1)}$	$\frac{1}{1.3s+1}$	0.0988	0.0988	1.3
P_3	$\frac{8.150(s^3+3s^2+3s+1)}{2s^3+5s^2+4s+1}$	$\frac{1}{2s+1}$	0.1226	0.1291	2

Table 2. Controller parameters.

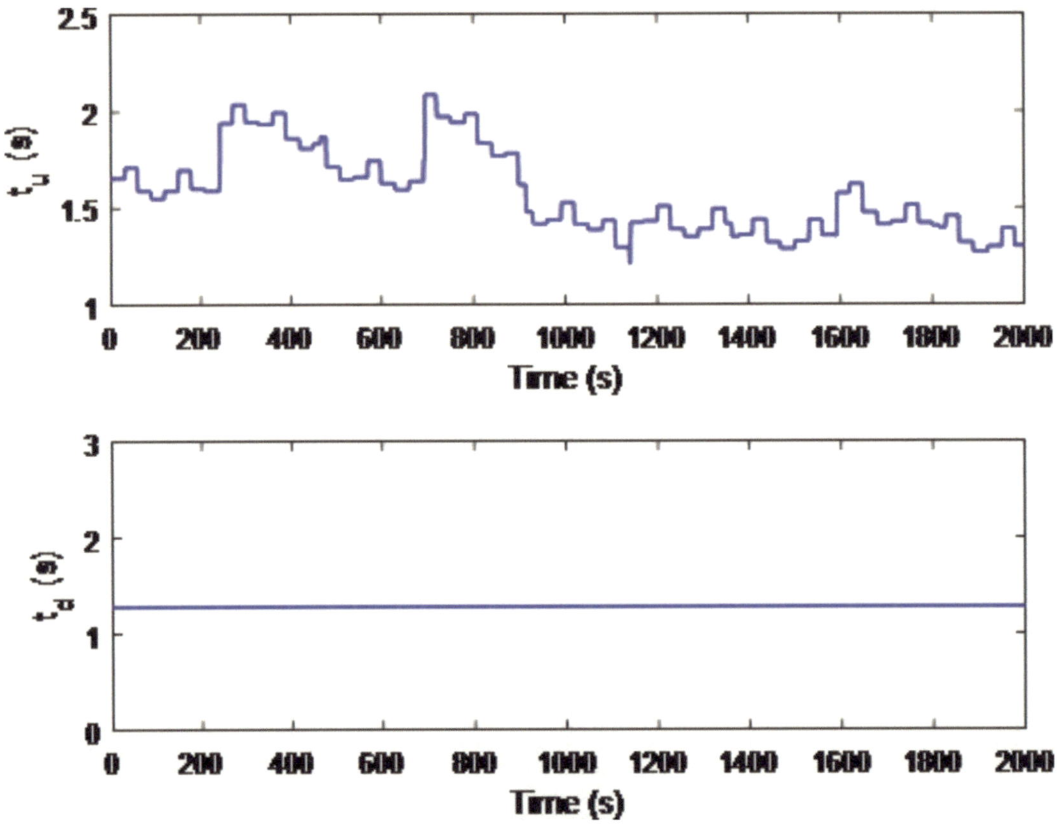

Figure 11. Network delay profile.

Delay type	Min	Max	Mean	Standard deviation
Upstream (s)	1.2140	2.0840	1.5734	0.2170
Downstream (s)	1.280	1.280	1.280	0.000

Table 3. Network delay statistics.

better than those of the PI controller. The numerical comparison assessed with respect to rise time (T_r), settling time before and after disturbance (T_{s1} and T_{s2}), overshoot (%OS), and integral time absolute error (ITAE) is given in **Table 4**. From the table, it is observed that the FASW produced less overshoot of 0.0284% compared to the respective 0.1938 and 4.1582% of SW and PI controllers, while the rise time and settling times of SW are shorter at 4.5980, 19.0756 and 185.5723 s, respectively, than those of FASW and PI.

It is worth noting that the initial control actions of SW and FASW are at 100%, while those of PI are at around 5%. This is due to the improvement of the setpoint weighting ability of the first two controllers.

http://dx.doi.org/10.5772/intechopen.70179

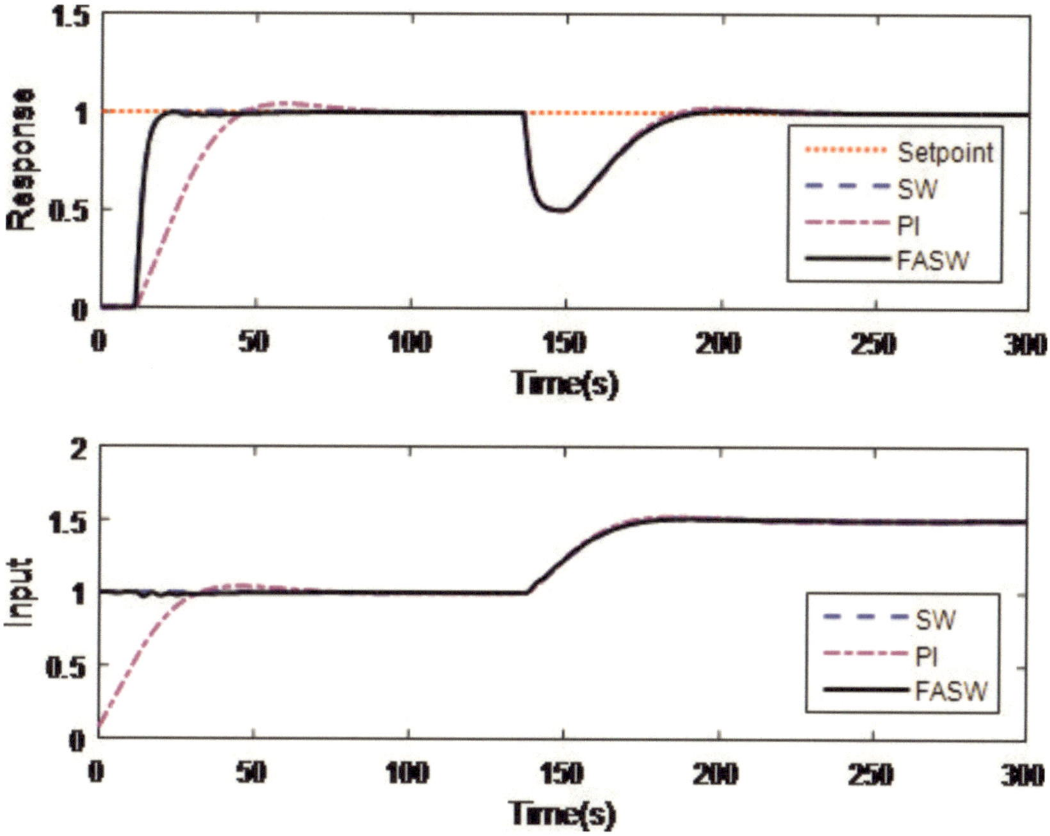

Figure 12. Response of first-order plant to load disturbance.

	T_r	T_{s1}	T_{s2}	%OS	ITAE
FASW	4.6129	19.5373	185.5723	0.0284	35.7358
SW	4.5980	19.0756	184.8150	0.1938	35.6524
PI	24.2732	76.1173	206.6751	4.1582	48.2429

Table 4. Performance of first-order plant.

To further evaluate the performance of the controllers, the plant is simulated to a variable setpoint signal and the result is shown in **Figure 13**. From the responses, it can be seen that during setpoint change both setpoint weighted controllers, i.e., FASW and SW, outperformed the PI controller.

4.2. Second-order plant

In a similar way to the first-order plant, the comparison of closed-loop response of this system for setpoint tracking and disturbance rejection with various controllers is shown in **Figure 14**

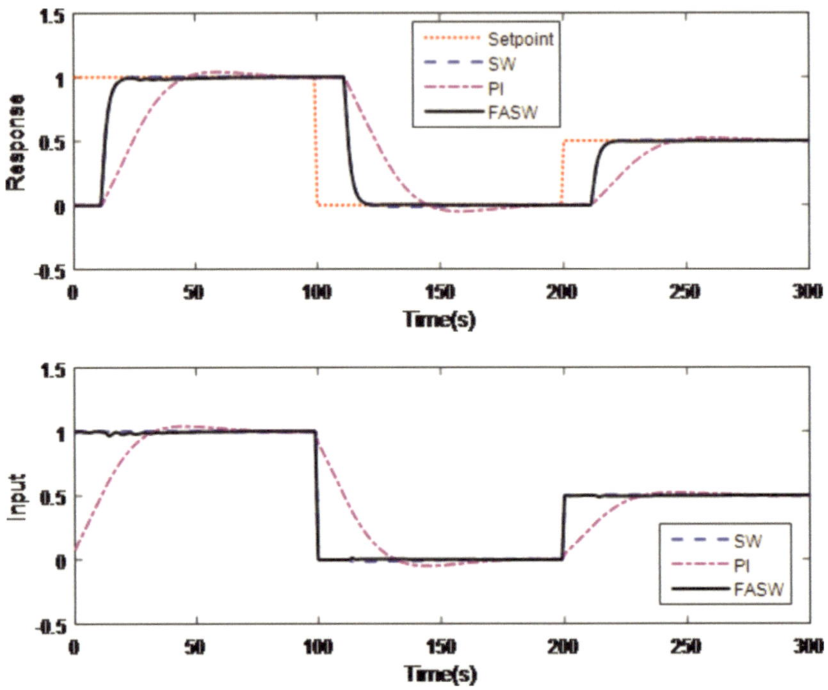

Figure 13. Response of first-order plant to changing setpoint.

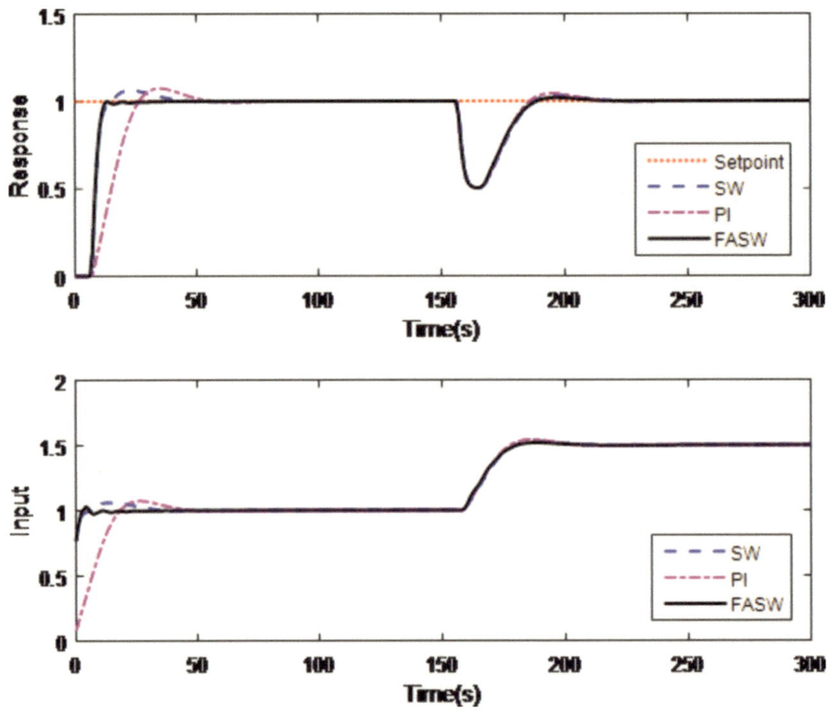

Figure 14. Response of second-order plant to load disturbance.

	T_r	T_{s1}	T_{s2}	%OS	ITAE
FASW	3.8653	11.8789	186.2306	0.0286	28.4034
SW	4.1330	37.1544	205.6308	6.1605	30.0163
PI	14.1246	49.2130	205.8253	7.3542	36.7180

Table 5. Performance of second-order plant.

and **Table 5**. From the figure, it is clearly seen that the FASW configuration achieved best tracking and disturbance rejection performance with least overshoot of 0.0286% compared to the 6.1605 and 7.3542% of the SW and PI, respectively. Furthermore, this configuration has the shortest rise and settling times for both before and after disturbance. The initial control signal of both SW and FASW is around 80% while that of the PI is around 10%. Furthermore, the comparison of variable setpoint tracking ability with various controllers is shown in **Figure 15**. From the responses, just as observed in the first-order plant, the tracking performance of FASW is better than that of SW and PI in terms of overshoot and undershoot during setpoint change.

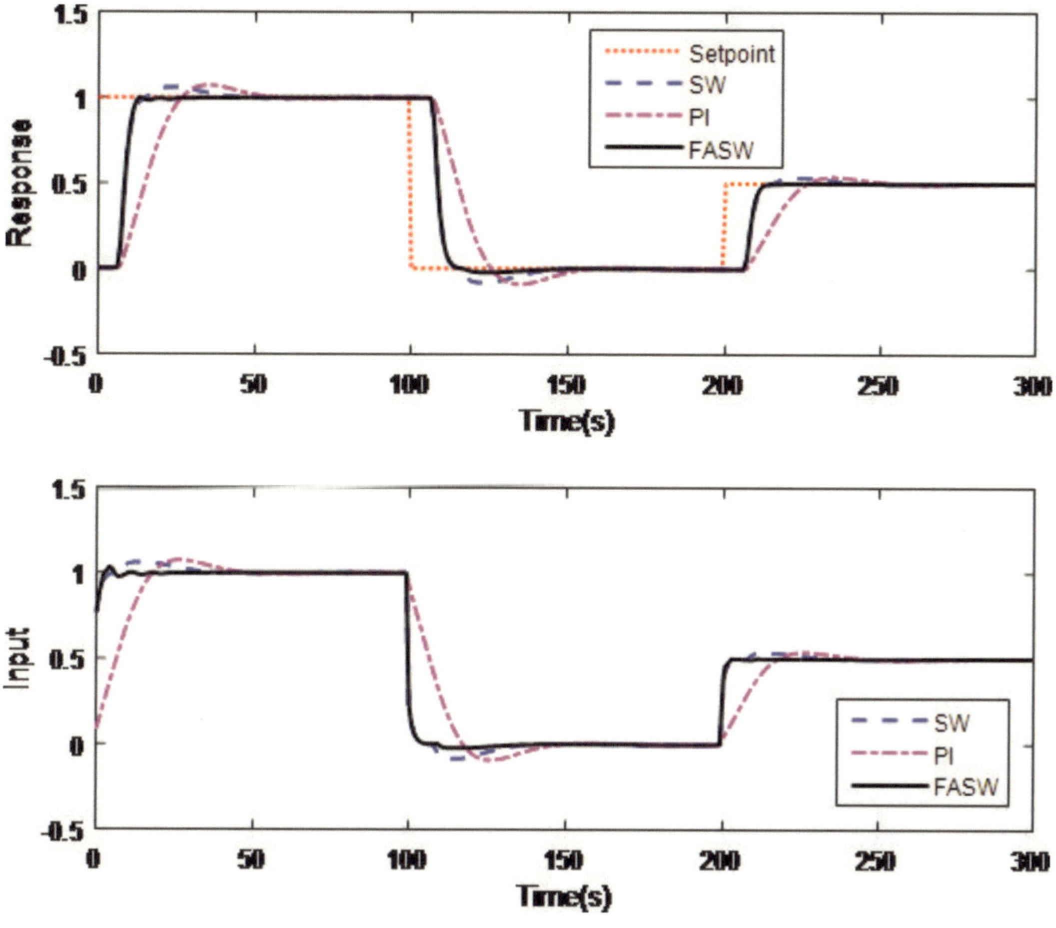

Figure 15. Response of second-order plant to changing setpoint.

4.3. Third-order plant

In a similar fashion to the earlier two plant models, the comparison of closed-loop response of the third-order system for setpoint tracking and disturbance rejection with various controllers is shown in **Figure 16** and **Table 6**. From both the figure and the table, it is clearly seen that the FASW configuration achieved best tracking and disturbance rejection performance with least overshoot 1.8137% as compared to the 9.3315 and 8.9940% of the SW and PI controllers, respectively. In addition, the proposed configuration has the shortest rise time of around 4.8 s compared to around 7.1 and 13.5 s of the SW and PI controllers. The settling times both before and after disturbance follow the same pattern. The two setpoint weighting configurations SW

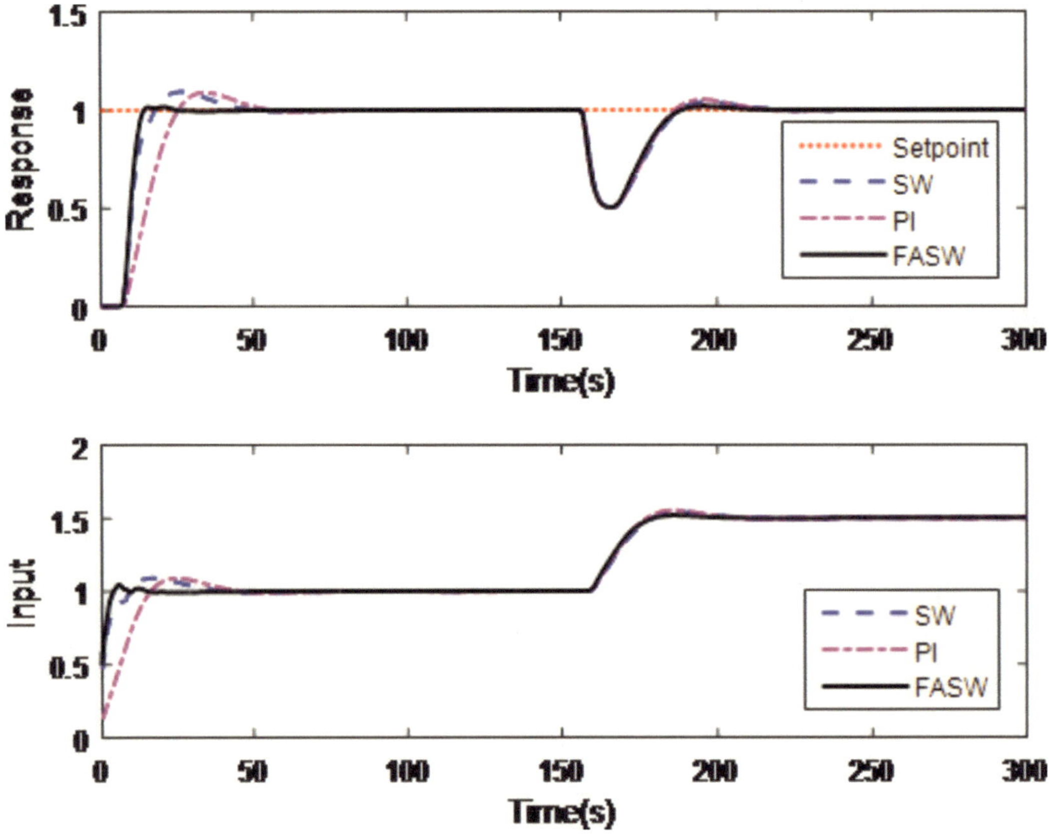

Figure 16. Response of third-order plant to load disturbance.

	T_r	T_{s1}	T_{s2}	%OS	ITAE
FASW	3.8653	11.8789	186.2306	0.0286	28.4034
SW	4.1330	37.1544	205.6308	6.1605	30.0163
PI	14.1246	49.2130	205.8253	7.3542	36.7180

Table 6. Performance of third-order plant.

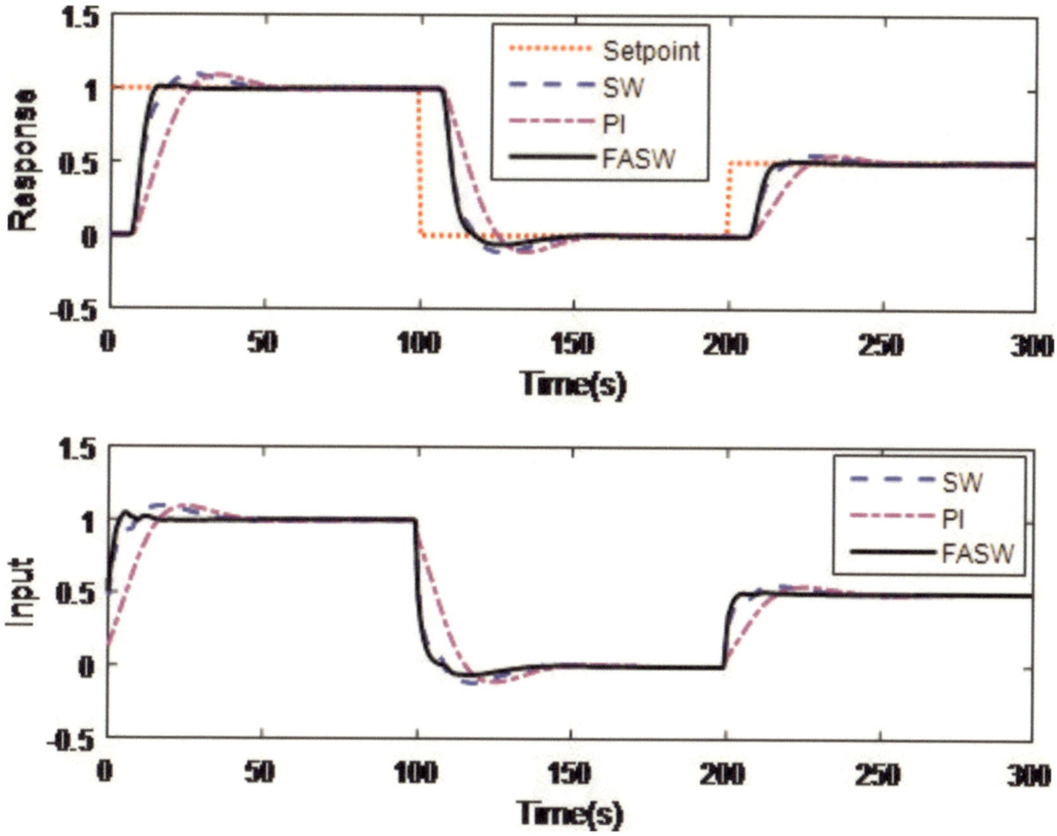

Figure 17. Response of third-order plant to changing setpoint.

and FASW as observed from the control signals are more aggressive than the PI controller at the beginning: starting at around 50% each.

The comparison of variable setpoint tracking ability with various controllers is shown in **Figure 17**. From the responses, it is seen that the tracking performance of FASW outperforms those of SW and PI. This is due to the adaptation ability of the FASW controller.

5. Conclusion

This chapter has presented an adaptation mechanism using fuzzy inference system for setpoint weighting controller designed for WirelessHART networked control environment. The adaptation mechanism adjusts the parameters of the setpoint weighting function at each sampling time. Result shows that the proposed approach is able to adapt the controller to variation in network delay. In comparison with ordinary PI controller and fixed setpoint weighting function, the adaptive mechanism has enabled significant improvement of the time domain performance of all the three plants considered. This is even more noticeable in the second- and third-order plants. Future work will focus on the implementation of the approach on a physical plant.

Author details

Sabo Miya Hassan, Rosdiazli Ibrahim*, Nordin Saad, Vijanth Sagayan Asirvadam, Kishore Bingi and Tran Duc Chung

*Address all correspondence to: rosdiazli@utp.edu.my

Department of Electrical and Electronic Engineering, Universiti Teknologi PETRONAS, Perak, Malaysia

References

[1] Zheng M, Liang W, Yu H, Xiao Y. Performance analysis of the industrial wireless networks standard: WIA-PA. Mobile Networks and Applications. 2015;**22**(1):1-12

[2] Petersen S, Carlsen S. WirelessHART versus ISA100. 11a: The format war hits the factory floor. IEEE Industrial Electronics Magazine. 2011;**5**(4):23-34

[3] Petersen S, Carlsen S. Performance evaluation of WirelessHART for factory automation. In: Proceedings of Emerging Technologies & Factory Automation; ETFA; September 22, 2009; IEEE; 2009. pp. 1-9

[4] Miya HS, Ibrahim RB, Saad NB, Asirvadam VS, Chung TD. WirelessHART process control with Smith predictor compensator. Advanced Science Letters. 2016;**22**(10):2676-2680

[5] Chen D, Nixon M, Mok A. Why WirelessHART. US: Springer; 2010. pp. 195-199

[6] Olenewa J. Guide to Wireless Communication. Cengage Learning; Boston, MA, USA 2013. pp. 185

[7] Hassan SM, Ibrahim R, Bingi K, Chung TD, Saad N. Application of wireless technology for control: A WirelessHART perspective. Procedia Computer Science. 2017;**105**:240-247

[8] Chung TD, Ibrahim RB, Asirvadam VS, Saad NB, Hassan SM. Adopting EWMA filter on a fast sampling wired link contention in WirelessHART control system. IEEE Transactions on Instrumentation and Measurement. 2016;**65**(4):836-845

[9] Blevins TL. PID advances in industrial control. IFAC Proceedings Volumes. 2012;**45**(3):23-28

[10] Blevins T, Chen D, Han S, Nixon M, Wojsznis W. Process control over real-time wireless sensor and actuator networks. In: Proceedings of High Performance Computing and Communications (HPCC), 2015 IEEE 7th International Symposium on Cyberspace Safety and Security (CSS), 2015 IEEE 12th International Conference on Embedded Software and Systems (ICESS), 2015 IEEE 17th International Conference; August 24, 2015; IEEE; 2015. pp. 1186-1191

[11] Tan KK, Tang KZ, Su Y, Lee TH, Hang CC. Deadtime compensation via setpoint variation. Journal of Process Control. 2010;**20**(7):848-859

[12] Lee TH, Tan KK, Tang KZ. Deadtime compensation via setpoint variation. In: Proceedings of 2009 Asian Control Conference; August 27, 2009; IEEE; 2009. pp. 512-517

[13] Hassan SM, Ibrahim R, Saad N, Asirvadam VS, Chung TD. Setpoint weighted WirelessHART networked control of process plant. In: Proceedings of Instrumentation and Measurement Technology Conference Proceedings (I2MTC); May 23, 2016; IEEE; pp. 1-6

[14] Gil P, Lucena C, Cardoso A, Palma LB. Gain tuning of fuzzy PID controllers for MIMO systems: A Performance-driven approach. IEEE Transactions on Fuzzy Systems. 2015;**23**(4):757-768

[15] Banu US, Uma G. Fuzzy gain scheduled CSTR with GA-based PID. Chemical Engineering Communications. 2008;**195**(10):1213-1226

[16] Dounis AI, Kofinas P, Alafodimos C, Tseles D. Adaptive fuzzy gain scheduling PID controller for maximum power point tracking of photovoltaic system. Renewable energy. 2013;**60**:202-214

[17] Khooban MH, Niknam T. A new intelligent online fuzzy tuning approach for multi-area load frequency control: Self Adaptive Modified Bat Algorithm. International Journal of Electrical Power and Energy Systems. 2015;**71**:254-261

[18] Chandrakala KV, Balamurugan S, Sankaranarayanan K. Variable structure fuzzy gain scheduling based load frequency controller for multi source multi area hydro thermal system. International Journal of Electrical Power and Energy Systems. 2013;**53**:375-381

[19] Sivanandam SN, Sumathi S, Deepa SN. Introduction to fuzzy logic using MATLAB. Berlin: Springer; 2007. pp. 387-391

[20] Esfahani HN, Azimirad V, Danesh M. A time delay controller included terminal sliding mode and fuzzy gain tuning for underwater vehicle-manipulator systems. Ocean Engineering. 2015;**107**:97-107

[21] Park KS, Ok SY. Fuzzy gain-tuning approach for active control system adaptable to physical constraints. KSCE Journal of Civil Engineering. 2015;**19**(5):1468-1474

[22] Radaideh SM, Hayajneh MT. A new fuzzy gain scheduling scheme for the PID controllers. Intelligent Automation and Soft Computing. 2003;**9**(4):267-277

Chapter 10

A Hybrid Sink Repositioning Technique for Data Gathering in Wireless Sensor Networks

Prerana Shrivastava

Additional information is available at the end of the chapter

http://dx.doi.org/10.5772/intechopen.70335

Abstract

Wireless sensor network (WSN) is a wireless network that consists of spatially distributed autonomous devices using sensors to cooperatively investigate physical or environmental conditions. WSN has a hundreds or thousands of nodes that can communicate with each other and pass data from one node to another. Energy can be supplied to sensor nodes by batteries only and they are configured in a harsh environment in which the batteries cannot be charged or recharged simply. Sensor nodes can be randomly installed and they autonomously organize themselves into a communication network. The main constraint in wireless sensor networks is limited energy supply at the sensor nodes so it is important to deploy the sink at a position with respect to the specific area which is the area of interest; which would result in minimization of energy consumption. Sink repositioning is very important in modern day wireless sensor network since repositioning the sink at regular interval of time can balance the traffic load thereby decreasing the failure rate of the real time packets. More attention needs to be given on the Sink repositioning methods in order to increase the efficiency of the network. Existing work on sink repositioning techniques in wireless sensor networks consider only static and mobile sink. Not much importance is given to the hybrid sink deployment techniques. Multiple sink deployment and sink mobility can be considered to perform sink repositioning. Precise information of the area being monitored is needed to offer an ideal solution by the sink deployment method but this method is not a realistic often. To reallocate the sink, its odd pattern of energy must be considered. In this chapter a hybrid sink repositioning technique is developed for wireless sensor network where static and mobile sinks are used to gather the data from the sensor nodes. The nodes with low residual energy and high data generation rate are categorized as urgent and the nodes with high residual energy and low data generation rate are categorized as non-urgent. Static sink located within the center of the network collects the data from the urgent nodes. A relay is selected for each urgent sensor based on their residual energy. The urgent sensor sends their data to the static sink through these relay. Mobile sink collects the data from the non-urgent sensors. The performance of the proposed technique is compared with mobile base station placement scheme mainly based on the performance according to the metrics such as average end-to-end delay, drop, average packet delivery ratio and average energy consumption.

Through the simulation results it is observed that the proposed hybrid sink repositioning technique reduces the energy hold problem and minimizes the buffer overflow problem thereby elongating the sensor network lifetime.

Keywords: wireless sensor networks, sink repositioning, energy efficiency, hybrid technique, network lifetime

---

1. Introduction

1.1. Wireless sensor network

A wireless sensor network (WSN) is a discrete network comprising of numerous wireless nodes referred to as sensors, which are deployed in order to perform the designated specific tasks like monitoring the surroundings and measuring the physical parameters such as temperature, pressure, humidity, etc. Since the location of an individual sensor cannot be preplanned or predetermined, these networks must have the potential to self-organize themselves. A wider geographic area can be covered by efficiently networking a large number of sensors thereby resulting in precise, dependable and robust networks. Wireless sensor networks are responsible to gauge, record, process and transfer the information to the destination node within the network zone using the assigned communication routes. Each sensor deployed in the network performs the functions like sensing the environment, processing the sensed data and communicating with the neighboring sensors. The sensor nodes have limited sensing range, processing power and energy levels.

The performance and efficiency of any wireless sensor network depends on the computational power, battery lifetime, data storage and communication bandwidth which in turn are directly dependent on the available energy levels. A major hurdle in the operation of sensors is the unavailability of an adequate energy. Normally the sensors depend upon their battery for power which in many cases cannot be replaced or recharged. Hence while designing any protocol for such networks, the conservation of the available energy of the sensor must be considered as an important factor. Thus, extending the lifetime of the sensor networks is a major area which is receiving a significant amount of interest from the research communities.

1.1.1. Structure of wireless sensor network

A basic sensor network consists of a large amount of sensor nodes. Each sensor is made up of small individual microcontroller fitted with sensors in which communication such as radios is used. The components of a sensor node are a sensing unit, a processing unit, a transceiver and a power unit. Generally, the sensor networks can form either a mesh topology or a star topology. Nodes can propagate by routing or flooding. In WSN, each node is assigned a number as its unique address for the purpose of communication. Functionally sensor nodes can be classified into two types. First, the nodes that deal within the network with other nodes and second, the ones which interface with the outside environment which are called as the gateway nodes or the sink nodes. The general structure of the wireless sensor network is shown in **Figure 1**.

As shown in **Figure 1**, the number of sensors is deployed in the geographical extent of the entire network and they will perform their task of sensing, processing, relaying and doing communication. All the information or the data that is sensed by the sensors will be forwarded to the sink node through multi hop relaying from where it will be provided to the end users.

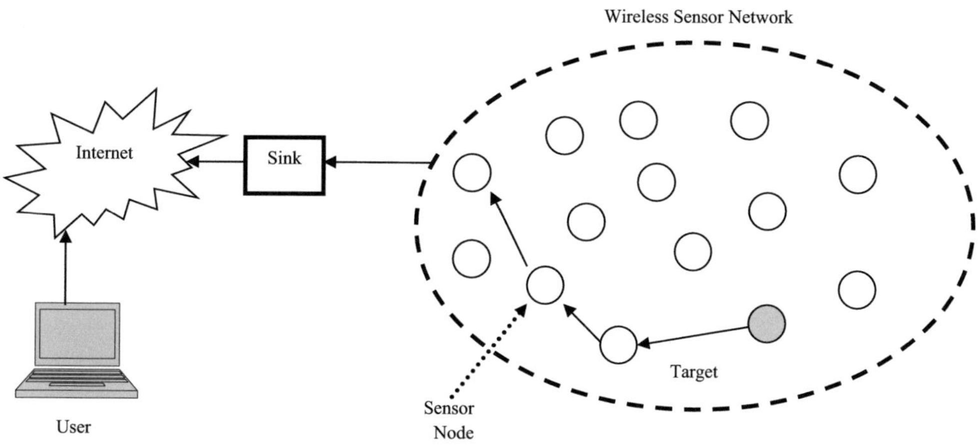

Figure 1. Structure of wireless sensor network.

1.1.2. Types of wireless sensor networks

Wireless sensor network typically has little or no infrastructure. There are two types of WSNs, namely structured model and an unstructured model. Structured model is deployed in a preplanned manner and it is used only for the network with fewer nodes. It has lower network maintenance and cost. Uncovered regions are not present in this model. Unstructured model is densely deployed in the network. The nodes that are deployed randomly have uncovered regions and are left unattended to perform the task. Maintenance is difficult here.

1.1.3. Characteristics of wireless sensor network

Some of the salient characteristics of the wireless sensor networks are described below:

- Dense sensor node deployment: typically, the sensor nodes are configured closely in WSN as compared to a Mobile Adhoc Network.
- Battery based energy source: energy can be supplied to the sensor nodes by batteries only and they are usually configured for harsh environment in which the batteries cannot be charged or recharged easily.

- Energy, computation and storage limitations: sensor nodes have limited energy, computation, and storage capabilities. Hence, the energy conservation measures are required in order to improve the efficiency and the life of the network.
- Self-configurable: generally, the sensor nodes can be randomly installed and they are capable of establishing a communication network by organizing themselves appropriately.
- Unreliable sensor nodes and data redundancy: sensor nodes are prone to physical damages or outages owing to their deployment in harsh or hostile conditions. The sensor nodes that are deployed close to each other play a similar role, in order to accomplish a common sensing task in a given area of concern. This results in building up of redundancy in events of failure.
- Application specific: depending upon the application, the design considerations of wireless sensor network will vary and may need customization.
- Frequent topology change: in most of the sensor network applications, the sensed data may pass through the various sensor nodes between source and the sink, thereby showing a many-to-one traffic pattern. The sensor node failure, damage, energy depletion, etc. may force the network topology to change continuously [1].

1.1.4. Major applications of wireless sensor networks

There are various applications of WSN that require constant monitoring and particular event detecting based on the requirement and features of the system. The applications can be divided into three categories [2] as mentioned in **Table 1**.

The importance of WSN is briefly described below for certain major applications as follows:

- Environmental and agricultural applications

WSNs are useful for the purpose of area monitoring, monitoring water levels as well as the rainfall. It is also used for forest fire & flood monitoring. Agricultural applications include sensing of chemicals, soil condition, irrigation planning, etc.

- Military applications

WSN's various characteristics are extremely useful in the area of enemy movement tracking, enemy intelligence information collection and transmission, surveillance, etc.

- Medical operations

Sensor networks play a critical role in monitoring the physiological readings of patient like blood pressure or pulse, etc. It plays an extremely significant role in post calamity medical relief operations such as earthquakes and floods.

- Heavy industrial monitoring

WSNs help in industrial applications by enabling to track material movement, warehousing, inventory planning and refurbishing, in spite of the harsh field conditions, consequently saving huge costs that are involved in such type of businesses.

Category	Examples
Applications requiring monitoring of physical entities	Urban Development and Planning; Medical Procedures and Healthcare Services, Robotics, etc.
Applications requiring monitoring of a geographical area	Ecological Studies, Space Research, Weather and Environmental studies, Agriculture, etc.
Applications requiring monitoring of both physical entities and geographical area	Military Applications, Wildlife Research, Disaster Management Operations, etc.

Table 1. Categories of application.

1.2. Sink repositioning

In WSN, sinks are bounded with abundant resources and the sensors that generate data are termed as sources. The sources can transmit data to a single or multiple sinks for the purpose of analysis and processing.

In wireless sensor networks, sink repositioning is preferred almost by all applications that involve real time communication. It helps to evenly distribute the traffic and hence minimize the packet loss or the data loss. To carry out sink repositioning, multiple sink deployment and sink mobility can be adopted. Precise information of the area being monitored is needed to offer an ideal solution by repositioning the sink.

1.2.1. Types of sink repositioning

Sink repositioning can be performed in the following ways.

Multiple sink deployment: in a given geographic area, multiple sinks can be deployed. By deploying multiple sinks in the network, the average number of hops through which the information has to pass through is decreased, since the data will always be sent to the nearest sink. Also by deploying multiple sinks, the load is evenly distributed among all the sinks [3].

Sink mobility: it is extremely advantageous in case of WSNs, if the sinks can move within the network boundaries with an acceptable delay. The mobile sink collects the data from the sensor nodes and also transmits it further. Although this approach results in comparatively higher time lag or latency, it helps in conserving the energy and hence increasing the life span [4].

Deploying multiple mobile sinks: multiple mobile sinks can be deployed in order to collect the data from the sensors in the given network without causing delay and buffer overflow problem. Here the mobile sink will relocate at regular intervals before the sensor's buffer overflows thereby avoiding the buffer overflow problem.

Initially, the research work in the field of wireless sensor networks mainly discussed the issues related to an uneven energy consumption which was leading to the energy hole problem in a sensor network. Generally, all the sensors generate data at a constant bit rate and transmit the data to the static sink through multihop transmission. Therefore the sensors which are closer to sink will die of energy soon, thereby creating an energy hole around the sink. The researchers have proposed an analytical modeling for the energy hole problem and using their model they have discussed the effectiveness of various techniques employed for justifying this problem.

2. Problem statement

During the regular network operation, relocating the sink is very challenging. During the sink's movement, the fundamental issues are when the sink should move, where the sink should move and how the data traffic would be handled when the sink is on the move. In a multi-hop network, finding an optimal location for the sink is very difficult. The difficulty mainly arises due to the following two factors. First, the sink can be moved to an infinite possible position. Secondly, a new multi-hop network topology needs to be established for every solution considered during the search for an optimal location [5].

Since employing the sink requires the precise knowledge of the monitored area, they are not always reasonable, even though the sink deployment can provide optimal solution. When accurate position of sensor is available and when nodes have motion capabilities, controlled deployment or online deployment is possible. The developing graph may have different properties during the online deployment. The basic issue in the sensor deployment is controlling the dynamic graph of mobile sensor networks [6]. The energy-unbalanced problem is another big challenge in sink deployment. Here the sensors that are closer to the sink are likely to consume their energy much faster than the other nodes [7] . When a network consists of multiple clusters, the relocation problem is significantly compounded. The sink cannot choose to move randomly around its cluster to enhance the intra-cluster network operation without considering the potential impact on inter sink connectivity that could impose on its capability to maintain communication with the sink nodes of other clusters [8]. Using the odd pattern of energy depletion, first the relocation of the sink has to be initiated even if it is considered as the most efficient network operation for a given traffic distribution at that time. The sink must make sure that no data is lost, when it is moving [9]. Using mobile sinks for data gathering has the drawback of buffer overflow problem. In other words, the sink has to visit each sensor node before its buffer overflows and this will depend on the speed of the mobile sink. However, it is very difficult to set the optimum speed for the mobile sink, since each sensor node has different buffer sizes and information generation rate. Apart from this problem, the residual energy of the sensors must also be considered, since sensors with low residual energy may deplete their energy before the mobile sink visits them.

3. Research issues addressed

In order to deal with the various issues in case of the wireless sensor network, the main objectives of the research is to design and implement a hybrid sink repositioning technique (HSRT) for data gathering in wireless sensor networks. The main focus has been on devising a technique which draws the benefits of both multiple sinks and sink repositioning, in order to improve the energy efficiency and various other performance metrics of the network. The design aspects of HSRT have been aimed at overcoming the energy hole problem and buffer overflow problem by taking into consideration the residual energy of the sensors that are deployed in the network.

4. Research methodology

4.1. Structure overview of hybrid sink repositioning technique (HSRT)

In the hybrid sink repositioning technique (HSRT), the sensors are randomly deployed within the geographic extent of the entire network. A single static sink and multiple mobile sinks are deployed in the network. The static sink is deployed at the center of the network. In case of sensors, the overflow of information occurs due to the limited storage capacity. The overflow time of each sensor is computed based upon their storage size and the data generation rate. All the sensors are then allotted a particular group based on their overflow time and location. After this one mobile sink is assigned to each group. Depending upon the data generation rate and residual energy of the sensors, the sensors are classified into two different categories namely urgent and non-urgent sensors. The static sink performs the function of collecting the data from the urgent sensors. A strategy has been devised in order to select and form the set of relay sensors, in such a manner that every individual urgent sensor has at least single relay sensor that is closest to the static sink. The urgent sensors transmits their information through the relay sensors to the final destination which is the static sink. In order to collect the data from the non-urgent sensors, a mobile sink deployment algorithm has been developed which will periodically collect the data from these sensors.

4.2. Sensor node classification

To explain the concept, a wireless sensor network with "i" number of sensors is considered. The sensor node classification has been done into two groups as urgent sensors and non-urgent sensors based on their residual energy and the data generation rate.

As shown in **Figure 2**, E_{ri} is the residual energy of the sensors, DG_{ri} is the data generation rate of the sensors, E_{rt} is the minimum threshold value of the residual energy and DG_{rt} is the maximum threshold value of the data generation rate. The sensors are classified as urgent and non-urgent sensors depending on the following two criteria.

If, $E_{ri} < E_{rt}$ and $DG_{ri} > DG_{rt}$ then the sensor is treated as urgent sensor.

Else if,

$E_{ri} > E_{rt}$ and $DG_{ri} < DG_{rt}$ then the sensor is treated as non-urgent sensor.

Thus a sensor having low residual energy and high data generation rate is categorized as urgent sensor and the sensor having high residual energy and low data generation rate is categorized as non-urgent sensors.

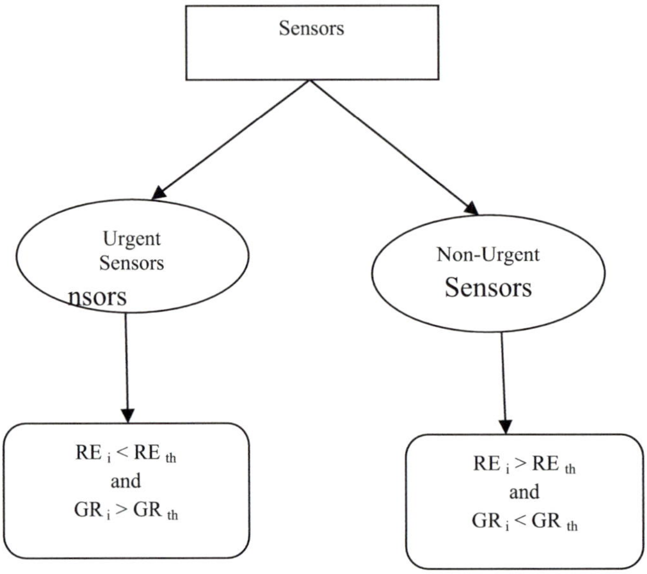

Figure 2. Classification of the sensors.

4.3. Positioning relay sensors near the static sink

For effective network operation and optimum performance, a two layer network is considered in a sensing field as shown in **Figure 3** wherein the relays and the static sink form the upper layer whereas the urgent sensors form the bottom layer.

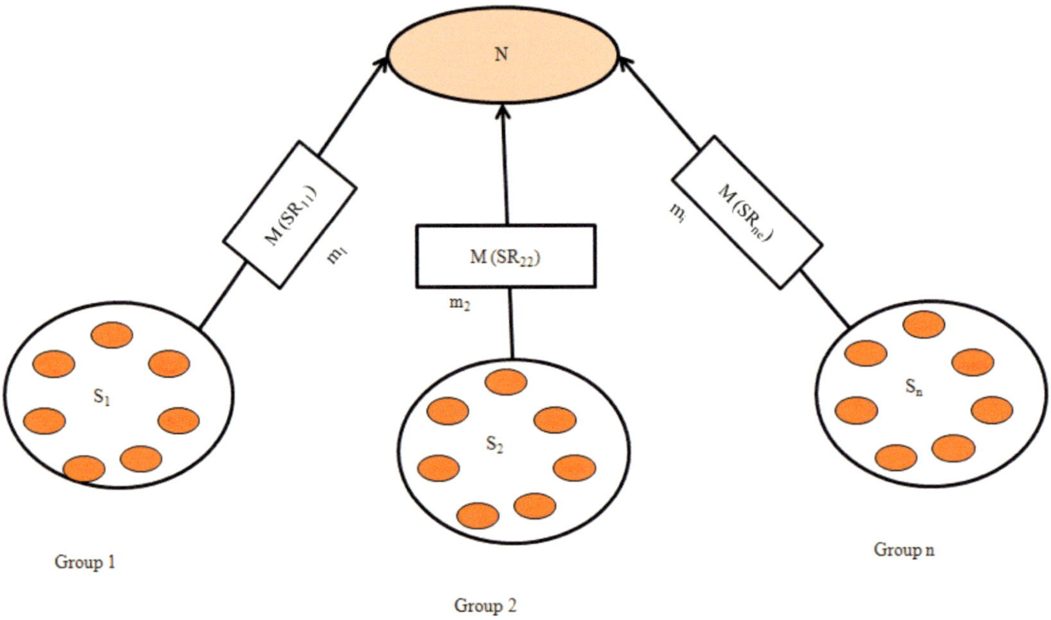

Figure 3. Formation of set of relay sensors.

Let,

N be the static sink,

$S = \{S_1, S_2, S_3 \ldots S_n\}$ be the set of urgent sensors,

$V = \{V_1, V_2, V_3 \ldots V_k\}$ be the set of non-urgent sensors and.

$R = \{R_1, R_2, R_3 \ldots R_e\}$ be the set of relay sensors.

In the given sensing field, the sensors are densely deployed whereas the relays are sparsely deployed. Data gathering is done by the joint co-operation of both sensors and the relays. A relay sensor is connected to the static sink in the upper layer of the network otherwise it is unconnected. Initially the set of the urgent sensors and the relays is not known as shown in **Figure 4**.

The main concern is to make use of the relay sensors having high residual amount of energy, in order to forward the information that is sensed by the urgent sensors to the static sink. A set of primary relays which are nearest connected relays to the urgent sensors S is determined. Let this set of relay sensors be denoted by M (SR_{ne}).The urgent sensors directs their data to M (SR_{ne}) and then M (SR_{ne}) relays this sensory data to the static sink N. In each interval the set of the relay sensors keeps on changing.

Now a set $H(m_i)$ is created such that,

$$H(m_i) = \{S_n \mid M(SR_{ne}) = m_i\} \tag{1}$$

where $H(m_i)$ is the set of all the urgent sensors attended by m_i. Each m_i will cover a set of all the relay sensors $M(SR_{ne})$ for all the urgent sensors S_n in different groups as depicted in **Figure 3**.

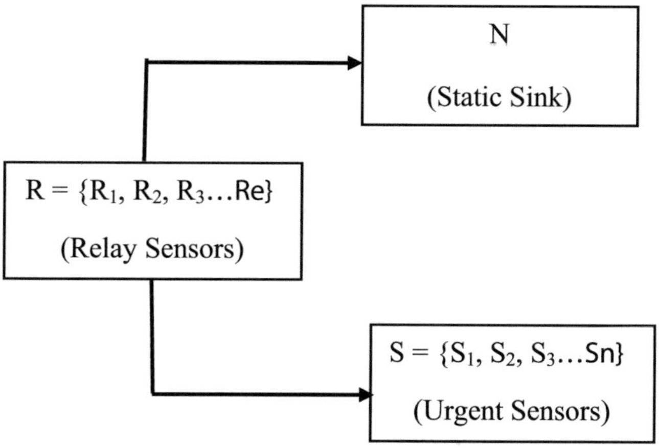

Figure 4. Two layer network.

4.4. Mobile sink deployment algorithm

A mobile sink deployment algorithm is developed in order to collect the data periodically from the non-urgent sensors. The main objective of this algorithm is to ensure that when the mobile sink is on the move it must travel minimum distance and at the same time perform maximum data collection.

All the mobile sinks, will first of all, identify those non-urgent sensors which are directly transmitting their information to them and at the same time the distance between these sensors and the mobile sink is less than the particular threshold value of the transmission distance. This is done because each mobile sink has its own capability of till what distance it can move while relocating. So a threshold value of the transmission distance for the mobile sink is selected. A set C_J is created where C_J denotes the set of the id's of those non-urgent sensors which are sending their information directly to the mobile sinks.

$$C_J = \{I = D_I^{(J)} < TD_{th}, \ I \in V\} \tag{2}$$

Once the set C_J is created, the mobile sink will wait for a particular duration during which each non-urgent sensor from C_J will transmit minimum one data packet to the mobile sink. The header of the data packet holds the ids of those sensors which are transmitting their information through these non-urgent sensors. As soon as the mobile sink receives the data packet from the non-urgent sensors from the set C_J, it records the ids of such sensors which are sending their data through these non-urgent sensors. Finally the mobile sink is able to identify the number of such sensors which are transmitting their own information through K, where, $K \in C_J$.

In order to reduce the mean distance between the non-urgent sensors and the mobile sink, the position of the distant sensors needs to be estimated. For this a set Z_k is created such that,

$$Z_k = \#\{I : k = \min D_I^{(k)}, k \in route_{IK}\} \tag{3}$$

Where Z_k is the set of the number of those distant sensors that transmit their information through the non-urgent sensors to the mobile sink and at the same time, the distance between them and the non-urgent sensors is minimum. Here $route_{IK}$ is the set of id's of the sensors on the route from sensor I to the non-urgent sensor k.

Once the mobile sink has identified that there are Z_k sensors communicating through non-urgent sensors k, the next task is to find the optimal position for the mobile sink. For this, the resultant route vector is used. The resultant route vector for sink j is approximated as,

$$RV_j = \frac{\sum_{k \in Q_j} U_k^{(j)} . Z_k}{\sum Z_k}, j = 1...K \tag{4}$$

where, RV_j is the Resultant route vector; $U_k^{(j)}$ is the unit vector from mobile sink j to the non-urgent sensor k; and Z_k is the set of number of distant sensors communicating through k.

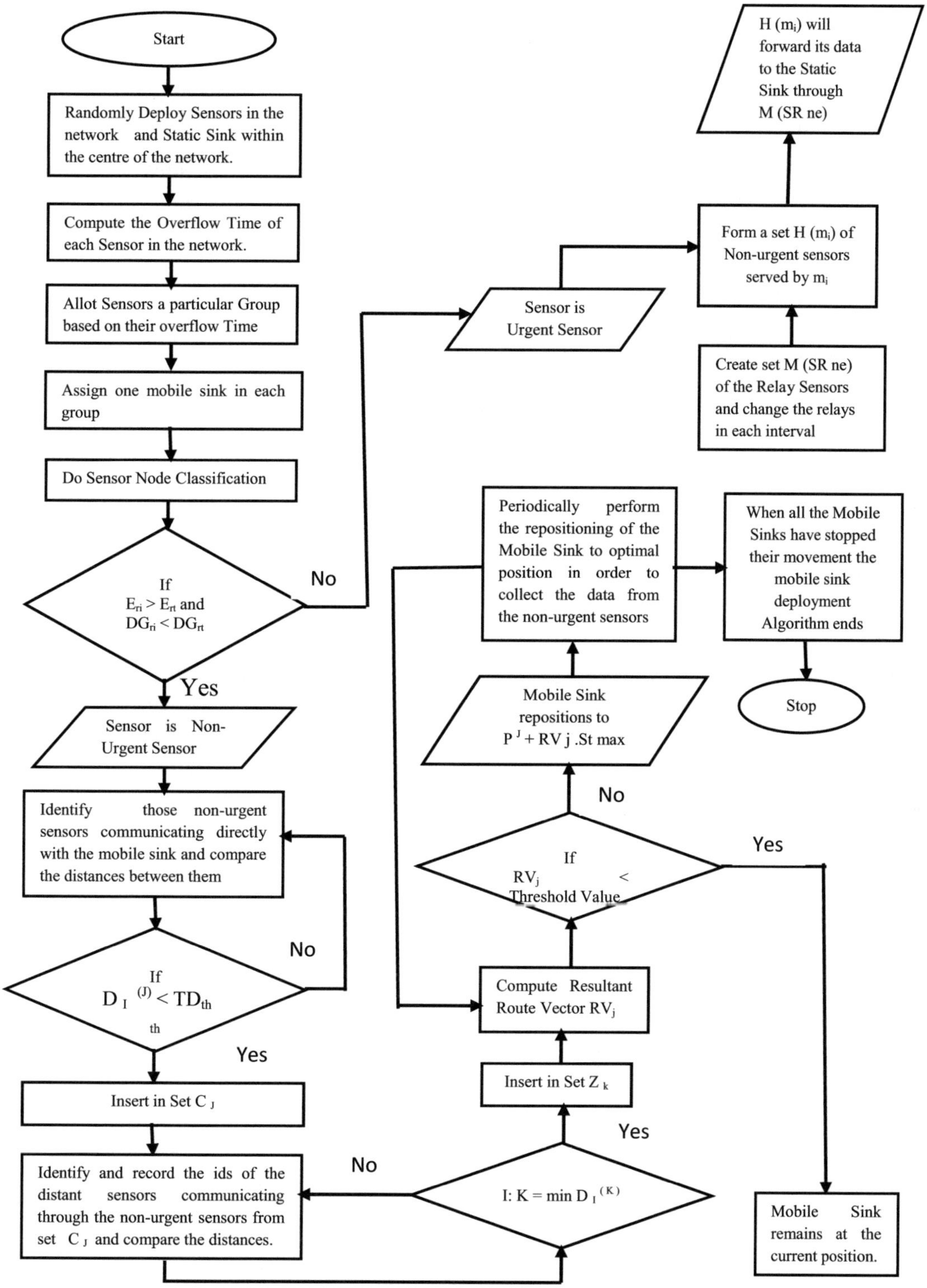

Figure 5. System flowchart of the HSRT Algorithm.

If the magnitude of the resultant route vector is less than a particular threshold value, then the mobile sink rests at its current position. On the other hand, if the resultant route vector is greater than a particular threshold value, then the mobile sink will reposition itself to a new location $P^J + RV_j . St_{max}$, where, P^J is the current position of the mobile sink and St is the maximum probable value of the stride that can be achieved by the mobile sink.

The process is repeated and the iteration continues, if the mobile sinks are moving for collecting the data from the non-urgent sensors. If all the mobile sinks have come to a standstill, then the mobile sink deployment algorithm terminates.

4.5. System flowchart of hybrid sink repositioning technique (HSRT)

The overall system flowchart of the hybrid sink repositioning technique (HSRT) that is designed for the purpose of data gathering in case of wireless sensor networks is depicted in **Figure 5**.

5. Research outputs and results

5.1. Simulation model and parameters

The implementation and the simulation of the hybrid sink repositioning technique (HSRT) are done by using the Network Simulator ns 2.32. A bounded region of 1000 × 1000 m^2 is considered in which the sensors are deployed using a rectangle distribution. The power levels are assigned to the sensors in such a way that their communication and sensing range is 250 m. In the simulation, the maximum data that can be supported by the communication media is fixed to 2 Mbps. The traffic generator used is the constant bit rate. The medium access control layer protocol used for the wireless local area network is the distributed coordination function of IEEE 802.11.

Table 2 depicts the various network parameters and their values which are assigned in the simulation model.

Area size	1000 × 1000 m^2
MAC	IEEE 802.11
Traffic source	CBR
Routing protocol	AODV
Simulation time	50 s
Packet size	500 Bits
Idle power	0.035 W
Transmit power	0.660 W
Receive power	0.395 W
Initial energy	10.1 J
Number of sensors	20, 40, 60, 80, 100
Rate	50, 100, 150, 200 and 250 kb

Table 2. Network parameters.

All the energy values have been selected based upon the energy model of ns2.32. Energy model represents the level of the energy in the sensors like the initial energy, idle energy and the usage of the energy for every packet it transmits and receives. The TCL script has been written for the HSRT. The NAM file is executed from the TCL script and it displays the network visualization of the HSRT. The NAM output which gives us the network visualization of HSRT is shown in **Figure 6**.

A single static sink as indicated by red color is deployed within the center of the network. The various sensors that are deployed in the network are assigned to a particular group. The multiple mobile sink as indicated by blue color are deployed in the network of the HSRT wherein each group is allotted one mobile sink, which will relocate itself inside the group that has been assigned to it, around every specific interval of time in order to collect the data from the non-urgent sensors. After running and executing the simulation, the mobile sink repositions itself to a new optimal location which is computed by the HSRT Algorithm, in order to collect the data from the non-urgent sensors of that particular group, as shown in **Figure 7**.

The number of sensors deployed in the network is increased and correspondingly the NAM output is observed before and after the sink repositioning by employing the HSRT, as shown in **Figures 8** and **9** respectively.

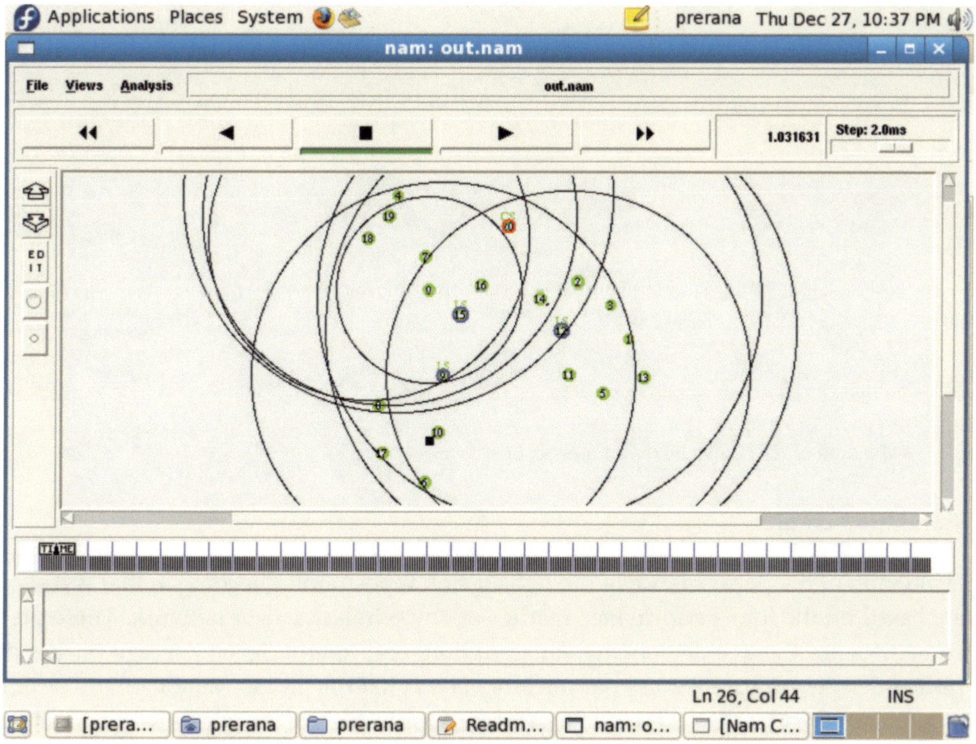

Figure 6. NAM output of HSRT.

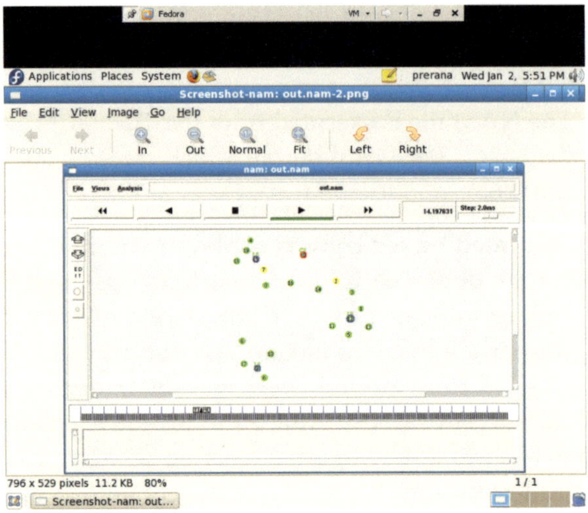

Figure 7. NAM output of HSRT after sink repositioning.

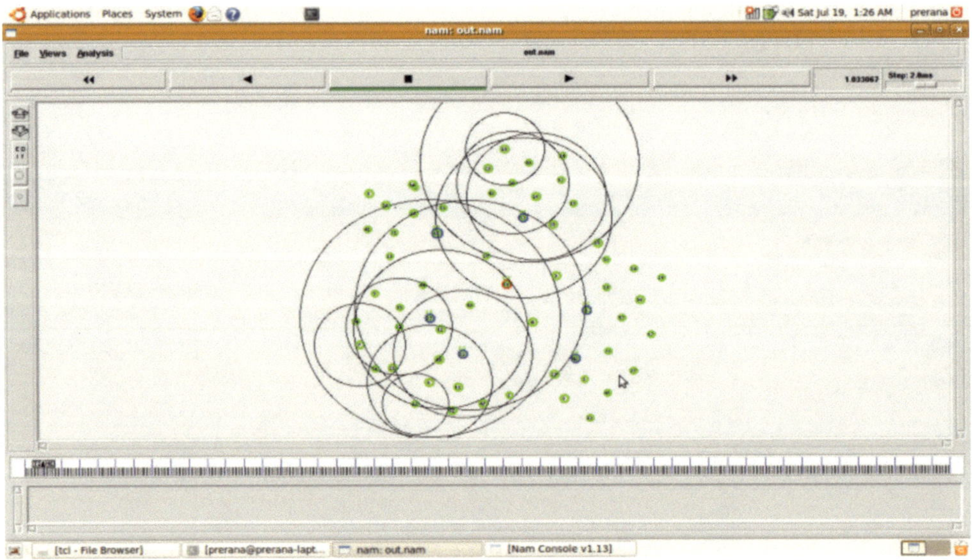

Figure 8. NAM output of HSRT with increased number of sensors.

5.2. Simulation results

The evaluation of the performance of the hybrid sink repositioning technique that is designed is done based on the four performance metrics of any wireless sensor network. These performance metrics are the average energy consumption, end to end delay, average packet drop and packet delivery ratio. All these parameters play a vital role in assessment of any designed technique, since the main focus is on data gathering application of the wireless sensor network.

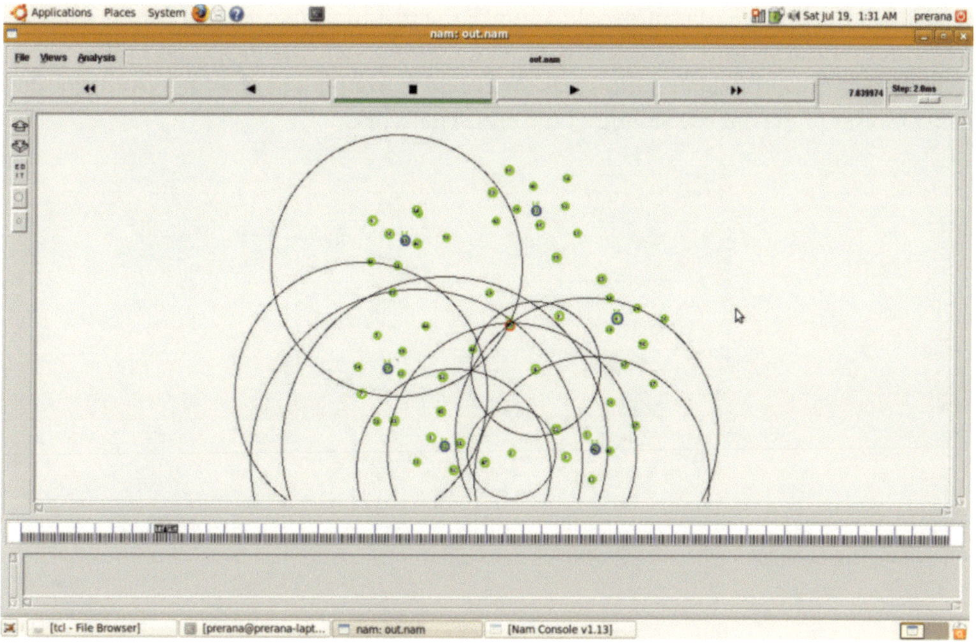

Figure 9. NAM output of HSRT after sink repositioning with increased number of sensors.

The performance of HSRT is compared with the existing multiple mobile base station placement scheme (MBSP) [10] for doing the necessary evaluation.

The tracing and monitoring of the simulation is done by running the TCL script which gives the trace values. The analysis of these trace values that has resulted from the simulation is done by making use of the trace data analyzer which is the X-Graph. The X-Graph is called within the OTCL script. The X-Graph will visually display the information of the trace values produced from the simulation.

The effect of HSRT on the various mentioned performance metrics is seen first by varying the number of sensors in the network and then by varying the speed of the mobile sinks.

5.2.1. Simulation results obtained by varying the number of sensors and the speed of the mobile sinks

In order to analyze the scalability of the HSRT, the number of sensors is varied from 20 to 100. The trace values for both HSRT and MBSP are monitored. **Figures 10–17** show the graphical representation of the simulation results obtained for various performance metrics by employing both HSRT and MBSP.

Figure 10 shows the average energy consumption for both the techniques, when the number of sensors is increased. The energy consumption increases almost linearly for the two techniques, when the network size is increased. It is observed that the HSRT consumes less energy when compared to the existing MBSP, since the relays are selected based on their residual energy. Moreover a particular threshold value of the residual energy is set for the sensors and therefore before the sensors completely deplete their energy, the proposed HSRT technique comes into picture and proper strategy as described is implemented which results in the significant amount of the energy saving of the entire network.

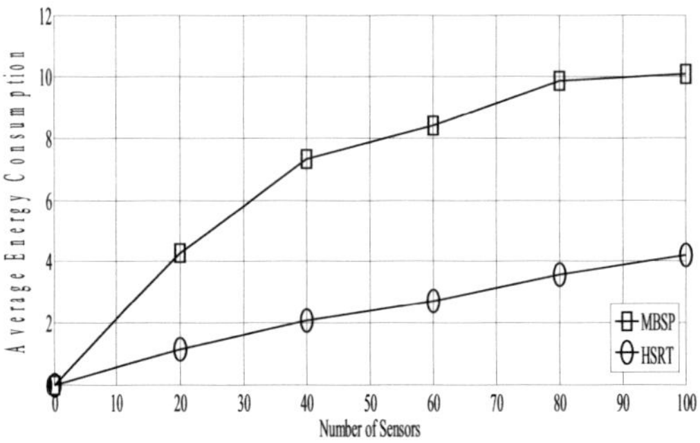

Figure 10. Sensors vs. average energy consumption.

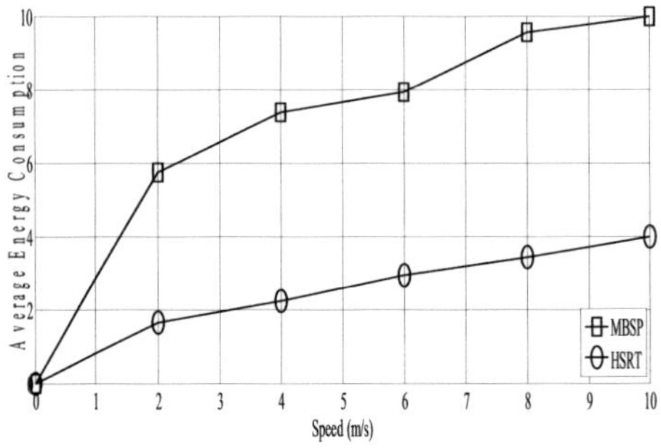

Figure 11. Speed vs. average energy consumption.

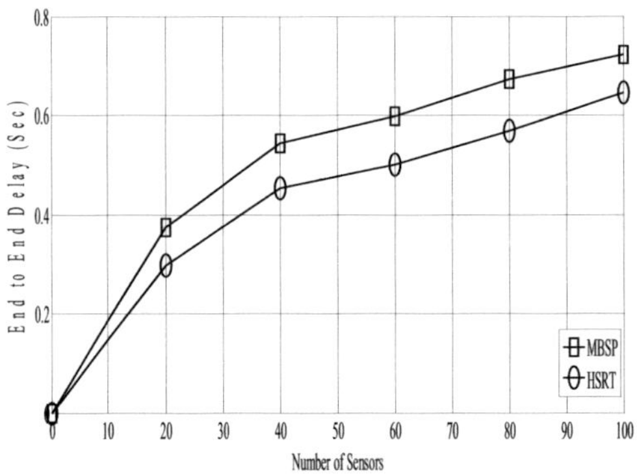

Figure 12. Sensors vs. end to end delay.

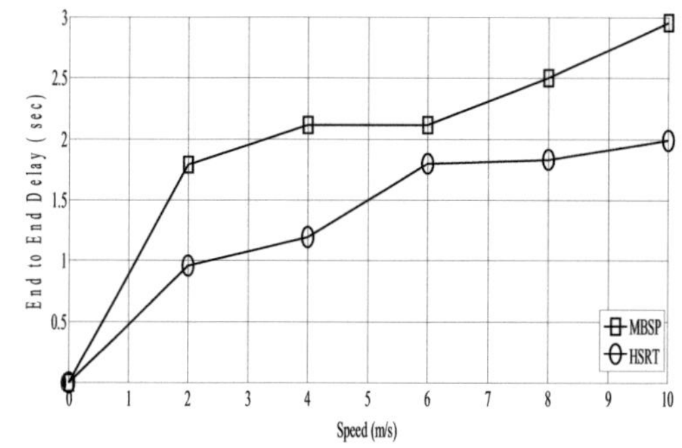

Figure 13. Speed vs. end to end delay.

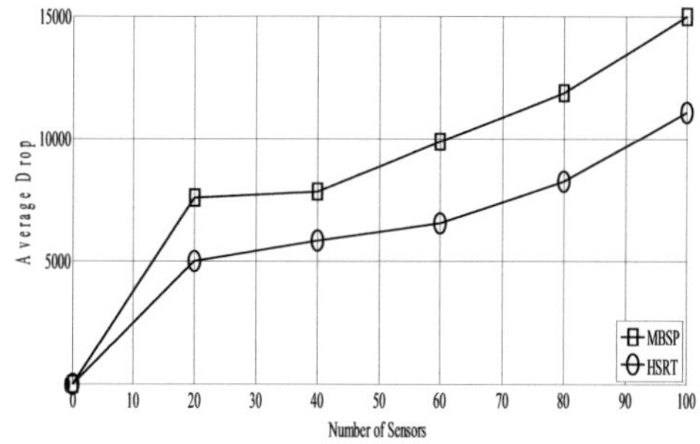

Figure 14. Sensors vs. average drop.

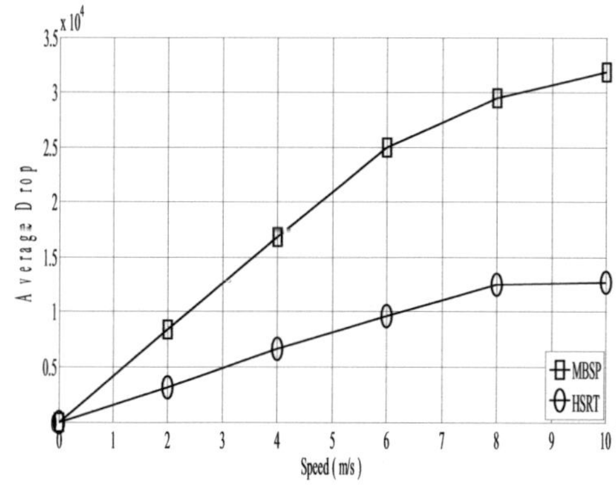

Figure 15. Speed vs. average drop.

In order to analyze the mobility of sinks, the speed of the mobile sinks is varied from 2 to 10 m/s. **Figure 11** shows the average energy consumption of both HSRT and MBSP when the speed of the mobile sink is increased. The energy consumption increases linearly as observed from the simulation. Moreover HSRT consumes less energy than MBSP.

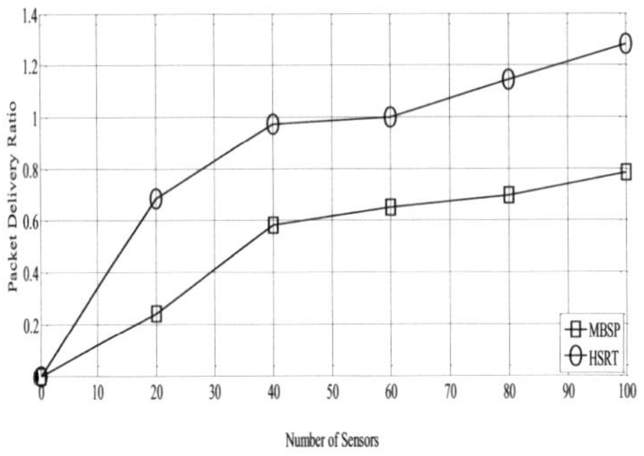

Figure 16. Sensors vs. packet delivery ratio.

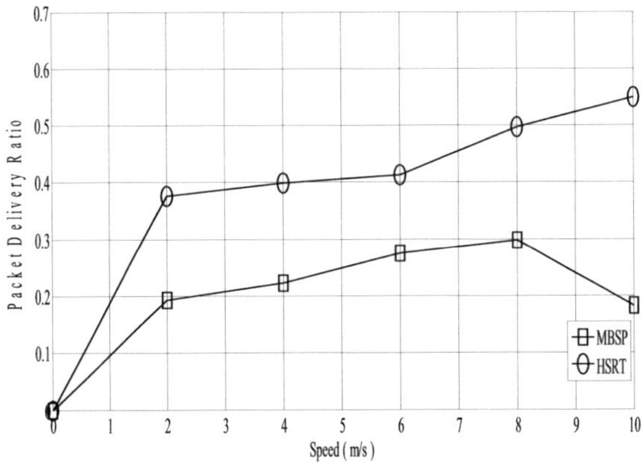

Figure 17. Speed vs. packet delivery ratio.

Figure 12 shows the average end-to-end delay in a scenario of varying number of sensors. When the network size is increased, it increases sink deployment time leading to the increased delay. From **Figure 12**, it is observed that HSRT minimizes the delay when compared with the existing MBSP scheme. In the proposed HSRT, the average number of hops that are involved in the transmission and reception of the data is minimized which also leads to the reduction of the overall end to end delay. Moreover the mobile sink is itself relocating at regular intervals to collect the data from the non-urgent sensors.

Figure 13 shows the results of average end-to-end delay when the speed of mobile sinks is increased from 2 to 10 m/s. It is observed that the delay increases beyond 0.3 seconds when the speed is above 6 m/s. The proposed HSRT shows a significant amount of improvement in end to end delay as compared to MBSP due to the proper distribution of the traffic load between the sinks as well as the sensors.

Figure 14 gives the average drop occurred for both the techniques when the number of sensors is increased. The increase in network size results in slight increase in packet drop. It can be seen that HSRT has less packet drop when compared to the existing MBSP, since all kind of losses that results from the energy hole problem and the buffer overflow problem are taken care of while designing the proposed hybrid sink repositioning Technique which results in the minimization of the packet drops.

On the other hand, **Figure 15** gives the results of packet drop for both the techniques when the speed of mobile sink is increased. When the mobile sink moves at higher speed, more buffer overflow will occur thereby resulting in more packet drops. This drawback is overcome in the proposed HSRT. Simulation results indicates that HSRT results in the reduction of the dropping of the data packets with the increasing speed of the mobile sinks.

Figure 16 shows the corresponding packet delivery ratio by varying the number of sensors. The increase in the network size results in the slight degradation of the delivery ratio. It is observed that HSRT achieves higher packet delivery ratio when compared with the existing MBSP technique. In HSRT the traffic load is evenly distributed among the sinks as well as the sensors. The use of mobile sinks which are relocating at regular intervals also decreases the number of hops. This ensures the enhancement in the delivery of the packets with less drop in the packets. The reduction in the Average Drop of the proposed HSRT gives rise to the improvement of the packet delivery ratio.

Figure 17 presents the packet delivery ratio when the speed of the mobile sink is increased. Higher the speed of the mobile sink larger will be the packet drops. Hence the delivery ratio decreases. But due to the optimum relocation of the sinks and data flow pattern, HSRT achieves higher delivery ratio than MBSP.

5.2.2. Percentage improvement of HSRT over existing MBSP

Table 3 shows the percentage improvement of HSRT when compared to the existing MBSP scheme.

Performance metrics	Algorithms	μ values obtained by varying the number of sensors	Percentage improvement	μ values obtained by varying the data rate	Percentage improvement
Average energy consumption	Existing MBSP	8.337066	30.44	11.526602	42.05
	Proposed HSRT	6.391499		8.114525	
End to end delay	Existing MBSP	0.5819653	18.07	2.2935636	47.36
	Proposed HSRT	0.4928986		1.5564357	
Average drop	Existing MBSP	10374.301	41.32	22294.65	87.35
	Proposed HSRT	7341		11,900	
Packet delivery ratio	Existing MBSP	1.4432616	41.98	0.6596911	47.66
	Proposed HSRT	1.0165246		0.4467636	

Table 3. Percentage improvement of HSRT.

6. Research conclusion

The hybrid sink repositioning technique is less complex and the overheads involved in running the algorithm is less and hence the proposed HSRT technique can be easily implemented in any real time applications like for the purpose of surveillance, military application or any other scenario where efficient data gathering is the prime focus and where each and every event needs to be detected properly.

Moreover the major hurdles directly affecting the performance of wireless sensor networks, namely the energy hole problem and the buffer overflow problem are minimized by the proposed hybrid sink repositioning technique that has been designed and successfully implemented. Through simulation results, it has been observed that employing the HSRT Algorithm enhances the overall functioning of the entire wireless sensor network in terms of the performance metrics namely the average energy consumption, end to end delay, average drop and packet delivery ratio. The improvement in all these performance metrics extends the lifetime as well as the accuracy of the WSN. Moreover HSRT also reduces the complexity involved in repositioning the multiple mobile sinks by employing the mobile sink deployment algorithm at regular intervals efficiently.

The research work presented in this chapter mainly focused on the energy consumption in terms of balancing and saving in order to extend the lifetime of the WSNs. Basically all the mobile sinks should remain active all the time in order to perform the task of data collection efficiently. But, also there are chances that they will remain idle most of the time if the data collection is light. The process of data collection gets affected or comes to a standstill if the mobile sink fails due to some fault. So the future scope or work must aim towards development of some effective visiting schedule and trajectory for the mobile sinks. Moreover it should include techniques for reducing the power consumption of idle sink and recovery of the failed mobile sinks.

Author details

Prerana Shrivastava

Address all correspondence to: prerana01@hotmail.com

Lokmanya Tilak College of Engineering, Navi Mumbai, India

References

[1] Coleri S, Puri A, Varaiya P. Power efficient system for sensor networks. Proceedings of 8th IEEE International Conference on Computers and Communication. July 2003

[2] Akkaya K, Younis M, Bangad M. Sink repositioning for enhanced performance in wireless sensor networks. Elsevier Journal on Computer Networks. 2005;**49**:512-534

http://dx.doi.org/10.5772/intechopen.70335

[3] Vincze Z, Vida R, Vidacs A. Deploying multiple sinks in multihop wireless sensor networks. Proceedings of IEEE Conference on Pervasive Services. July 2007

[4] Younis M, Akkaya K. Strategies and techniques for node placement in wireless sensor networks: A survey. Elsevier Journal on Ad Hoc Networks. 2008;**6**:621-655

[5] Lit X-Y, Xu XH, Wang SG, Tang SJ, Dait G, Ji Zhao Z, Qi Y. Efficient data aggregation in multi hop wireless sensor networks under physical interference model. IEEE. 2009

[6] Anastasi G, Conti M, Di Francesco M, Passarella A. Energy conservation in wireless sensor networks: A survey. Elsevier Journal on Ad Hoc Networks. 2009;**7**:537-568

[7] Mollanejad A, Mohammad Khanli L, Zeynali M. Dynamic station repositioning using genetic algorithms in wireless sensor networks. International Journal of Computer Sciences Issues. March 2010;**7**(2):24-29

[8] Yang T, Mino G, Barolli L, Durresi A, Xhafa F. Energy saving in wireless sensor networks considering mobile sensor nodes. Proceedings of IEEE International Conference on Complex, Intelligent and Software Intensive Systems. April 2011:249-256

[9] Li X, Yan S, Xu C, Nayak A, Stojmenovic I. Localized delay bounded and energy efficient data aggregation in wireless sensor networks. International Conference on Wireless Communication and Mobile Computing. 2011

[10] Wu F-J, Hsu H-C, Tseng Y-C. Non-location-based mobile sensor relocation in a hybrid static-mobile wireless sensor network. Proceedings of 3rd IEEE International Conference on Sensor Technologies and Applications, IEEE Computer Society. 2009